KB133079

Gelato

SORBETTO
GRANITA
COLD DESSERT

Gelato

SORBETTO, GRANITA, COLD DESSERT
젤라또, 소르베또, 그라니따, 콜드 디저트

초판 1쇄 발행 2023년 9월 1일
초판 2쇄 발행 2024년 7월 5일

지은이 유시연 | **영문 번역** 김예성 | **펴낸이** 박윤선 | **발행처** (주)더테이블

기획·편집 박윤선 | **디자인** 김보라 | **사진** 조원석 | **스타일링** 이화영
영업·마케팅 김남권, 조용훈, 문성빈 | **경영지원** 김효선, 이정민

주소 경기도 부천시 조마루로385번길 122 삼보테크노타워 2002호
홈페이지 www.icoxpublish.com | **쇼핑몰** www.baek2.kr (백두도서쇼핑몰) | **인스타그램** @thetable_book
이메일 thetable_book@naver.com | **전화** 032) 674-5685 | **팩스** 032) 676-5685
등록 2022년 8월 4일 제 386-2022-000050 호 | **ISBN** 979-11-92855-01-1 (13590)

• 이 책에서 사용하는 외래어는 국립국어원이 정한 외래어 표기법에 따르나, 일부 단어는 이탈리아어의 발음에 가깝게 표기했습니다.

더 테이블
THE TABLE

젤라또
소르베또
그라니따
콜드 디저트

Gelato

SORBETTO
GRANITA
COLD DESSERT

YOO SIYEON

유시연

더 테이블
THE TABLE

PROLOGUE

1

'이민 가방을 들고 이탈리아 밀라노로 떠나다.'

초등학교 졸업 후 떠난 이탈리아 밀라노.

나는 밀라노에서 중학교와 고등학교를 다니며 미술과 건축을 공부하였다. 만약 내가 한국으로 귀국하지 않았다면 아마 그곳에서 대학교를 진학하여 건축학을 전공했을 것이다.

어린 나에게 젤라또는 학교가 끝나고 집으로 돌아가는 길에 당연히 손에 하나 들고 가던 일상 속의 작은 기쁨이었다.

한국으로 돌아와 대학교를 진학하였고, 유학 시절 나의 일상이었던 젤라또가 잊힐 때쯤, 그 나이 누구나 하는 진로 고민을 하게 되었다.

2

'신문 스크랩 한 장으로 시작된 젤라또 인생'

나는 어릴 때부터 회사 생활을 하겠다는 생각보다는 나의 일을 하고 싶다는 생각이 강했다. 친구들이 취업 준비를 할 때도 난 그저 어떤 기술을 가지면 내가 즐기며 살아갈 수 있을지에 대한 고민을 했던 것 같다. 그러던 어느 날, 엄마가 건내준 신문 스크랩 한 장으로 나의 미래는 확고해졌다.

3

'다시 이탈리아로 떠나자!'

앞으로 몇 년 뒤 한국에는 콜드 디저트 시장이 열릴 것이라는 어떤 셰프님의 인터뷰 기사를 읽었었다. 아이스크림, 젤라또, 그라니따 등의 이야기를 다룬 신문 기사가 이탈리아 생활 중 소소한 기쁨이었던 젤라또를 다시 떠올리게 하였다.

대학 졸업 후 다시 시작한 이탈리아 생활은 어린 날과는 달랐다. 나는 이탈리아 전 지역을 돌아다니며 자유롭게 젤라또를 배웠다. 젤라또 문화의 시작인 이탈리아 중부 피렌체부터 북부의 밀라노, 남부의 시칠리아, 마지막으로 칼피지아니 젤라또 대학교Carpigiani Gelato University가 있는 볼로냐까지.

우리나라의 김치가 지역마다 특색이 다르듯 젤라또 역시 이탈리아 내에서 지역마다 다른 특색을 보인다. 그래서 각 지역을 대표하는 젤라또 교육기관을 다녔고, 그들이 생각하고 즐기는 이탈리아 젤라또 문화를 온전히 느끼고 배울 수 있었다.

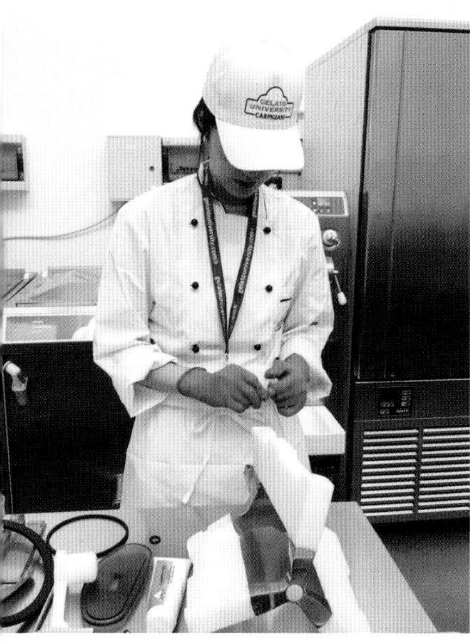

'Leaving for Milan, Italy with a Large Suitcase'

After graduating from elementary school, I left for Milan, Italy.

I went to middle and high school in Milan, where I studied art and architecture. If I hadn't returned to Korea, I would have majored in architecture at a university there.

To me, as a child, a cup of gelato was a small joy in my daily life that was a natural routine on my way home from school.

I returned to Korea and went to a university. By the time the memory of gelato was diminishing, which was a routine while studying abroad, I started to think about the career everyone at that age would do.

'The Beginning of Gelato Life by a Newspaper Clip'

From an early age, I had a strong will to do my own work rather than work in a company. Even when my friends were preparing for employment, I was just thinking about what kind of skills I would need to enjoy living with. Then one day, my future was solidified with a piece of newspaper clipping my mom gave me.

'I'm Going Back to Italy!'

I read one chef's interview article who said that a cold dessert market would open in Korea in the next few years. The newspaper article about ice cream, gelato, granita, and so on reminded me of gelato, which was a small bliss in Italian life.

Life in Italy after graduating from university was different from my childhood days. I traveled all over Italy and learned gelato liberally from Florence in central Italy, where gelato culture began, to Milan in the north, Sicily in the south, and finally to Bologna, where Carpigiani Gelato University is located.

Just as Korean kimchi has different characteristics from region to region, gelato also has different distinctions within Italy. So, I went to a gelato institution representing each region, and I was able to fully appreciate and learn about the Italian gelato culture that they think of and enjoy.

4

'일쀼il più 연구소를 오픈하고 다시 공부를 하다.'

젤라토를 처음 배울 때의 계획은 한국에서 젤라떼리아를 오픈하기 위해서였지만 막상 이탈리아에서 배워온 문화와 지식과 기술을 한국에 적용시키기에는 사용되는 재료나 국내에서 대중화되어 있는 맛의 차이 등 여러 가지 상황이 달라 많은 연구가 필요했다. 그래서 젤라또 제조에 필요한 모든 장비를 구입한 후 매장이 아닌 나의 첫 공간 '일쀼 연구소'를 오픈했다. 연구소에서 한국의 재료와 그에 맞는 레시피 연구를 하며 나에게 부족한 것이 무엇인지 고민에 빠졌다. 그 고민의 결과로 젤라또를 더 잘 이해하기 위해서는 한국의 외식문화와 업계 트렌드를 알아야 한다는 생각이 들었다. 그래서 대학원에 진학하여 석사와 박사 과정을 밟으며 학문적 지식을 쌓고 있다. 기술적인 부족함으로는 젤라또 맛을 개발하고 만들기 위해서 제과 지식과 기술을 더 알면 도움이 될 것이라는 판단에 2019년, 하던 일과 학업을 모두 멈추고 프랑스 파리로 떠났다. 파리의 유명한 여러 제과 교육기관 중 에꼴 벨루에 꽁세이Ecole Bellouet Conseil를 다니며 짧은 시간이나마 프랑스 제과 문화와 기술을 배우고 다양한 재료를 조합하는 기술을 폭넓게 사용할 수 있게 되었다.

5

'Carpigiani Gelato University 공식 한국 교육자'

2016년 겨울, 전 세계에서 가장 크고 유명한 젤라또 교육기관인 Carpigiani Gelato University에서 한국을 담당하는 젤라또 교육자가 되어줄 수 있겠냐는 제안이 들어왔다. 나를 더 성장시킬 수 있는 좋은 기회였기에 흔쾌히 받아들였고, 한국과 이탈리아를 오가며 젤라또를 교육하는 사람이 되었다. 2017년부터는 나와 공부한 사람들이 손으로 꼽을 수 없을 만큼 많아졌고, 이제는 한국 젤라또 업계에서는 '유시연'하면 'Carpigiani', 'Carpigiani'하면 '유시연'이라고 대입되며 해외 기사에서는 '한국의 젤라또 여왕'이라는 과분한 타이틀로 나를 소개하기도 하였다.

'Opening of il più Lab to Start Studying Again'

The original plan to learn gelato was to open a gelateria in Korea. However, to apply the culture, knowledge, and skills learned in Italy to Korea, I needed to research more due to various circumstances, such as the difference in ingredients used and the taste popularized in Korea. Hence, after purchasing all the equipment needed for gelato, I opened my first workroom, 'il più Lab.' At the lab, I researched Korean ingredients and suitable recipes and had to think about what I lacked. As a result, I figured that to understand gelato better, I needed to know Korean dining culture and industry trends. I decided to attend graduate school to pursue my master's and doctoral degrees to build my academic knowledge. And I realized that knowing more confectionery knowledge and techniques would be helpful due to a lack of technical skills. So, in 2019, to further develop and make gelato flavors, I stopped all my work and studies and left for Paris, France. While attending Ecole Bellouet Conseil, one of the well-known pastry institutions in Paris, I was able to learn the French pastry culture and techniques and use them to combine various ingredients widely, although it was a short period of time.

'Official Korean Instructor of Carpigiani Gelato University'

In the winter of 2016, I was offered to become a gelato instructor in charge of Korea at Carpigiani Gelato University, the world's largest and most famous gelato institution. It was an excellent opportunity to develop myself better, so I accepted it readily and became an instructor, traveling back and forth between Korea and Italy. Since 2017, the number of people who studied with me increased so much that it was countless. And now, in the Korean gelato industry, they refer to 'Siyeon Yoo' as 'Carpigiani,' and 'Carpigiani' as 'Siyeon Yoo,' and overseas articles introduced me to the overwhelming title of 'Korea's Gelato Queen.'

6

'㈜젤라또 코리아 HACCP 인증 제조공장 오픈'

젤라또 교육과 컨설팅을 하면서 내가 만든 젤라또를 소비자에게 판매하여 반응을 보고 싶은 생각이 생겼다. 그래서 2020년 일뿌 연구소를 정리하고 HACCP 인증 제조 공장인 ㈜젤라또코리아를 오픈하였다. 현재는 5성급 호텔, 유명 브랜드 카페, 고메 마켓, 고급 레스토랑 등에 젤라또와 소프트 젤라또 액상 믹스를 납품하고 있으며 업체에서 원하는 다양한 맛과 형태로 ODM 생산을 하는 전문 회사로 성장했다.

7

'대한민국 최초 젤라또 월드컵 본선 진출'

이탈리아 '리미니'라는 도시에서는 매년 1월 '씨젭SIGEP'이라는 전시회가 열린다. 젤라띠에레라면 꼭 한 번, 아니 그 이상 가봐야 하는 중요한 전시회인데, 이 전시회에서 '젤라또 월드컵' 대회가 2년에 한 번씩 개최된다. 2015년부터 SIGEP 전시회 대회장에서 각국의 대표 선수들을 보며 언젠가는 나도 저 자리에서 내 기술을 전 세계인에게 평가받겠다는 꿈을 꾸었고, 이 꿈은 2024년 1월에 현실이 된다. 작년 10월 싱가포르에서 열린 아시아 젤라또 컵에 단장으로서 2명의 선수와 함께 참가하여 본선 진출권을 획득하였다. 대한민국 최초로 진출한 본선 무대에는 새로운 선수 2명을 추가 영입하여 총 4명의 선수와 함께 참가한다. '젤라또 월드컵' 대회는 각 지역의 예선 대회를 통과한 12개국을 대표하는 선수들이 젤라또와 젤라또 케이크, 젤라또 플레이팅 디저트, 초콜릿 공예 및 아이스 카빙 등 다양한 기술들을 선보이는 유일한 세계대회다. (한국에서도 많은 관심과 응원을 받는 대회가 되기를 바란다.)

8

'이 책의 독자 분들에게'

이탈리아에서 어렵게 열심히 배웠던 젤라또를 한국 시장에 맞게 적용시켜온 나의 시간들이 이 책에 고스란히 담겨 있다. 젤라또의 기본 개념, 레시피 작성 방법, 다양한 젤라또, 소르베또, 그라니따 레시피 등 젤라띠에레 뿐만이 아닌 콜드 디저트에 관심 있는 모든 이들에게 도움이 될 수 있도록 최대한 쉽게 풀어냈고 예시를 들어 자세하게 설명했다.

이 책 한 권으로 여러분들이 걸어갈 젤라띠에레의 길이 나보다는 조금 덜 험난하고, 덜 힘들기를 바란다. 이 책이 여러분의 첫 걸음을 잘 뗄 수 있는 좋은 길잡이가 될 수 있었으면 좋겠다.

'Opening of Gealto Korea Co., Ltd. HACCP Certified Manufacturing Plant'

While instructing and consulting gelato, the idea of selling my own gelato to consumers and seeing their reactions came across my mind. So, in 2020, I closed my il più lab and opened Gelato Korea Co., Ltd., a HACCP certified manufacturing plant. Currently, we supply gelato and soft gelato liquid mix to 5-star hotels, famous brand cafes, gourmet markets, and high-end restaurants. We have grown into a specialized company that produces ODM in various flavors and shapes requested by the customers.

'Korea's First Gelato World Cup Finalist'

In the city of Rimini, Italy, an exhibition called 'SIGEP' is held every January. If you are a gelatiere, it is an important exhibition that you must visit at least once, and the 'Gelato World Cup' competition is held at this exhibition every two years. Since 2015, watching the national players of each country at the SIGEP exhibition venue, I dreamed of the day I would have my skills evaluated by people around the world, which soon becomes a reality in January 2024. In October of last year, I participated in the Asia Gelato Cup held in Singapore with two players as a team leader and won the final round. In the final stage, which is the first in Korea, two new players are additionally recruited to participate, with a total of four players. The 'Gelato World Cup' competition is the only global competition in which players representing 12 countries that have passed the preliminaries in each region showcase various techniques such as gelato, gelato cake, gelato plated dessert, chocolate craft, and ice carving. (I hope this competition can receive much attention and support in Korea.)

'To The Readers'

This book contains my efforts in adapting gelato, which I learned hard and diligently in Italy, to suit the Korean market. It explains the basic concept of gelato, the recipe formulating method, various gelato, sorbetto, granita recipes, etc., in detail with examples to help not only gelatiere but also those interested in cold desserts.

I hope that with this book, the path to becoming a gelatiere will be less rugged and challenging than mine. I pray this book can be a good guide to your first steps.

저는 대한민국의 젤라띠에레, 유시연 입니다.
I am Gelatiere from Korea, Siyeon Yoo.

CONTENTS

Part

Understanding Gelato

젤라또 이해하기

Part

Gelato Making Machines and Tools

젤라또를 만드는 기계와 도구들

Part

Ingredients Used in Gelato

젤라또에 사용되는 재료들

Part

Theory for Writing Gelato Recipes

젤라또 레시피를 작성하기 위한 이론

Part 5

Gelato Recipes

젤라또 레시피

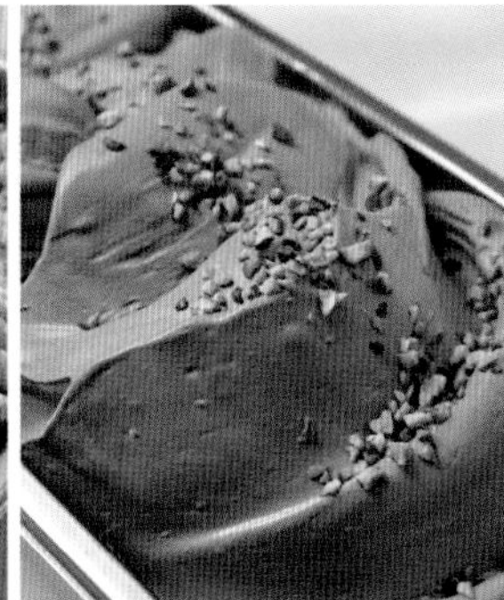

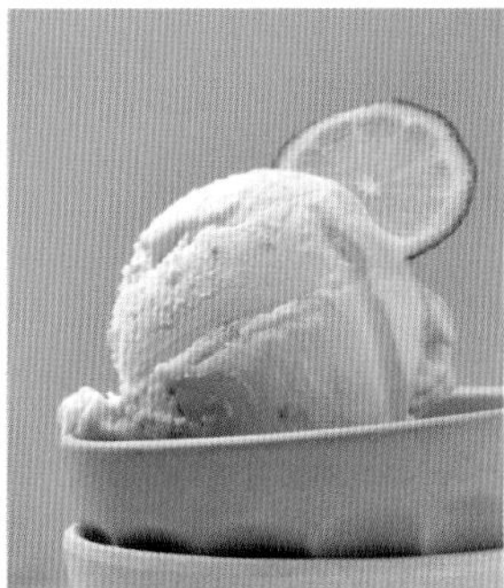

dherm

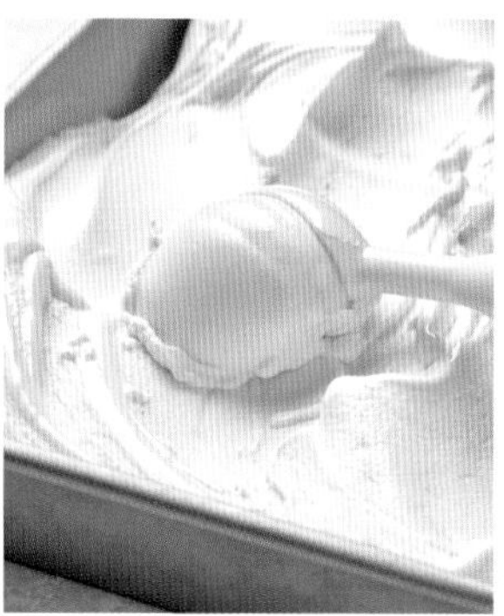

Part

Theory for Writing Sorbetto Recipes

소르베또 레시피를 작성하기 위한 이론

Part

Sorbetto Recipes
소르베또 레시피

Part

Theory for Writing Granita Recipes

그라니따 레시피를 작성하기 위한 이론

Part

Granita Recipes

그라니따 레시피

Part

Cold Dessert Recipes

콜드 디저트 레시피

INDEX

Part 1.

Understanding Gelato

젤라또 이해하기

- 젤라떼리아 : 젤라또를 판매하는 곳
- 젤라띠에레 : 자신만의 레시피로 젤라또를 제조하는 사람

- Gelateria : A place that sells gelato
- Gelatiere : A person who makes gelato with one's own recipe

1 What is Gelato?

젤라또란?

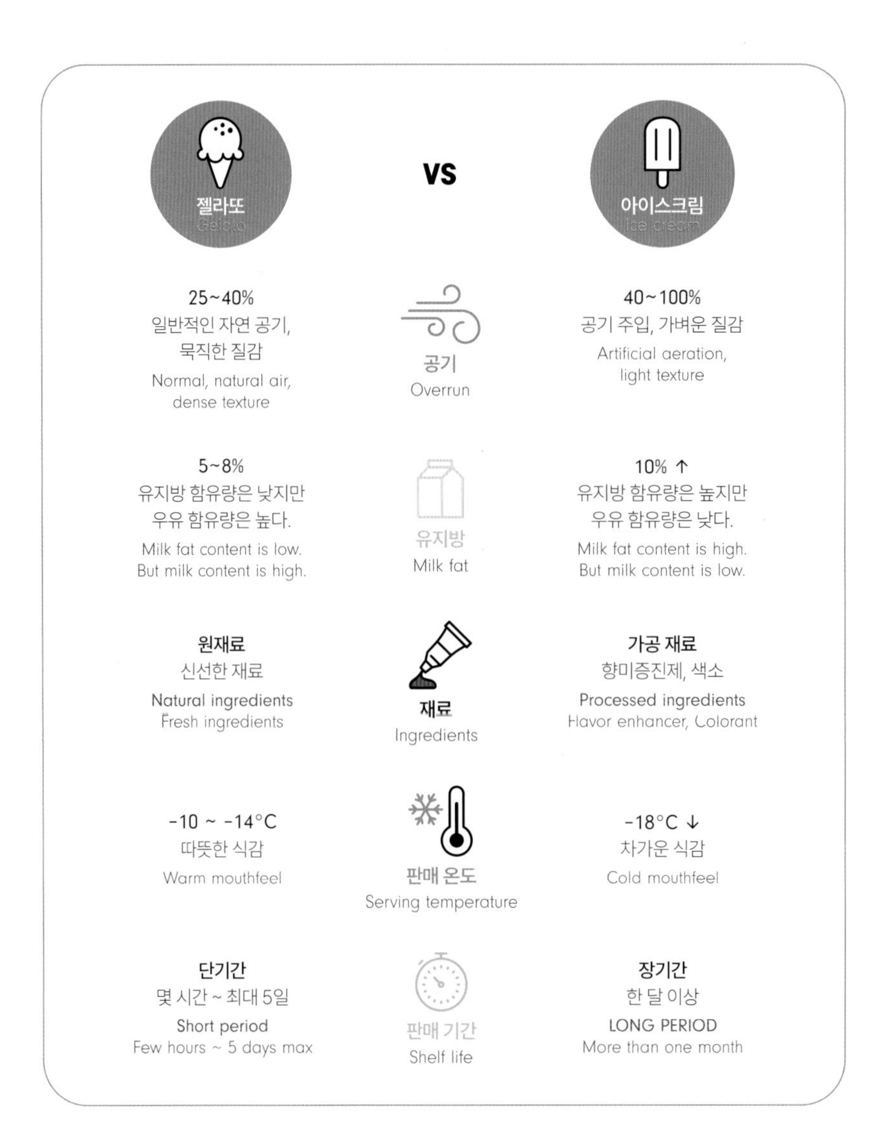

젤라또
Gelato
VS
아이스크림
Ice cream
25~40%
일반적인 자연 공기,
묵직한 질감
Normal, natural air,
dense texture
공기
Overrun
40~100%
공기 주입, 가벼운 질감
Artificial aeration,
light texture
5~8%
유지방 함유량은 낮지만
우유 함유량은 높다.
Milk fat content is low.
But milk content is high.
유지방
Milk fat
10% ↑
유지방 함유량은 높지만
우유 함유량은 낮다.
Milk fat content is high.
But milk content is low.
원재료
신선한 재료
Natural ingredients
Fresh ingredients
재료
Ingredients
가공 재료
향미증진제, 색소
Processed ingredients
Flavor enhancer, Colorant
-10 ~ -14°C
따뜻한 식감
Warm mouthfeel
판매 온도
Serving temperature
-18°C ↓
차가운 식감
Cold mouthfeel
단기간
몇 시간 ~ 최대 5일
Short period
Few hours ~ 5 days max
판매 기간
Shelf life
장기간
한 달 이상
LONG PERIOD
More than one month

젤라또는 아이스크림과 다르게 밀도가 높고 부드러운 질감과 재료 본연의 맛이 있는 그대로 느껴지는 이탈리아 대표 콜드 디저트이다.

젤라또와 아이스크림의 가장 큰 차이점은 흔히들 오버런이라고 이야기한다. 하지만 일반 소비자들은 그 의미가 쉽게 다가오지 않을 것이다. 아이스크림은 인위적으로 공기를 주입시켜 부피를 크게 키우고 부드러운 질감을 만들기 위해 지방 함량을 높이지만, 젤라또는 인위적으로 공기를 주입시키지 않으며 지방 함량도 낮고 그 안에서 부드러움을 유지하는 것이 특징이다.

Unlike ice cream, gelato is a symbolic cold dessert in Italy that has a dense, soft texture and tastes the original ingredients as they are.

The most significant difference between gelato and ice cream is often said to be the overrun. However, general customers will come to that meaning slowly. Ice cream is aerated artificially to increase its volume and increase fat content to create a smooth texture. On the other hand, gelato is not artificially aerated and has a low-fat content but maintains its softness in it.

소비자 관점에서는 젤라또와 아이스크림은 서빙 온도, 즉 입 안에서 느껴지는 온도와 맛의 조합이 가장 직접적으로 와닿는 차이점일 것이다.

From the customer's point of view, the difference between gelato and ice cream would be the serving temperature, that is, the temperature felt in the mouth and the combination of flavors.

젤라또의 서빙 온도는 -10 ~ -14°C 정도이며 가장 이상적인 서빙 온도는 -12°C이다. 하지만 계절, 날씨, 쇼케이스가 설치된 공간의 환경, 젤라또 레시피에 따라 여러 가지 변수들이 발생할 수 있으므로 젤라띠에레가 서빙 온도 및 기타 사항을 직접 컨트롤할 수 있어야 한다.

Serving temperature of gelato is between -10~-14°C, and the most ideal serving temperature is -12°C. However, various variables can occur depending on the weather, the environment of the space where the showcase is installed, and recipe of gelato. Therefore, the gelatiere should be able to directly control the serving temperature and other matters.

아이스크림의 경우 -18°C 이하의 온도에서 서빙되므로 젤라또보다 더 차갑고 단단하게 느껴지며, 아이스크림 스쿱을 사용해 동그란 모양을 깔끔하게 만들어낼 수 있다.

Ice cream is served at a temperature of -18°C or lower, so it feels colder and harder than gelato, and you can use an ice cream scoop to make a clean round shape.

✦ 젤라떼리아에서 아이스크림 스쿱을 사용하고 싶은 경우 -12°C보다 온도를 낮춰야 서빙하기가 편하다. -12°C에서 아이스크림 스쿱을 사용할 경우 젤라또의 질감이 부드러워 딱 떨어지는 동그란 모양을 만들기가 쉽지 않다.

✦ If you want to use an ice cream scoop in a gelateria, it is easier to serve if the temperature is lower than -12°C. If you use an ice cream scoop at -12°C, it is not easy to make a neat round shape due to the gelato's soft texture.

이탈리아 젤라떼리아의 90% 이상은 사용하는 주재료의 맛이 곧 최종적인 맛으로 완성된다. 예를 들어 피스타치오를 주재료로 선택하면 입 안에서 느껴지는 젤라또의 맛도 직관적인 피스타치오 맛이다. 그렇기 때문에 젤라또의 이름 역시 주재료가 제품명이 되는 경우가 많다.

In terms of flavor, the main ingredient used determines the final taste in more than 90% of Italian gelateria. For example, if pistachio is selected as the main ingredient, the flavor of gelato felt in the mouth is also intuitively pistachio. For this reason, the name of gelato is often the name of the main ingredient.

반면 아이스크림은 젤라또처럼 주재료가 최종적인 맛으로 완성되는 경우보다 토핑류나 소스 및 다양한 재료를 혼합하여 복합적인 맛으로 만드는 경우가 많다. 이에 따라 아이스크림의 이름 역시 새로운 이름을 만들어 이름만 보고는 무슨 맛인지 알 수 없는 경우가 많다. (예: 엄마는 외계인, 할머니의 레시피, 골드 메달 리본, 러브 포션 등)

On the other hand, ice cream is sometimes made with the main ingredient as the final flavor, like gelato. But in many cases, it's made by mixing various ingredients and using toppings or sauces to create a complex flavor. As a result, the name of the ice cream is given a creative name, so it is often difficult to know what flavor it is just from the name. (e.g., Mother is an alien, Grandma's recipe, Gold medal ribbon, Love potion, etc.)

젤라또와 아이스크림의 당 함량은 비슷하지만, 지방 함량이 젤라또가 더 낮아 아이스크림보다 칼로리가 낮다. 일반적으로 미국에서 생산되는 아이스크림은 지방 함량이 기본적으로 10%이며 20~25%에 이르기도 한다. 반면 젤라또는 젤라띠에레마다 선호하는 지방 비율은 다르지만 보통 5~12% 정도이다. 결과적으로 지방 함량이 높은 아이스크림은 재료의 맛을 표현하기 위해 양산형 제품의 경우 인공향료를 사용하여 지방의 풍미를 뚫고 나올 수 있게 맛을 연출한다. 반면 젤라또는 지방 함량이 높지 않아 원재료 자체만을 사용해도 충분히 그 맛을 연출할 수 있다.

Gelato and ice cream have similar sugar content, but because gelato has less fat content, it has lower calories than ice cream. Generally, ice creams in the United States start with a basic fat content of 10% and can reach 20~25%. On the other hand, the preferred content of fat varies by gelatiere, but it is usually between 5~12%. As a result, ice cream with a high-fat content uses artificial flavors in the case of mass production to show the taste of the ingredients by breaking through the flavor of the fat. Meanwhile, gelato has no high-fat content, so you can express the flavor enough by using only the raw ingredients.

매장에서 만드는 수제 아이스크림은 만드는 방법과 사용되는 재료가 젤라또와 동일하지만, 서빙 온도를 더 낮게 판매하거나 토핑류를 많이 혼합하여 제조하는 경우가 많다. 수제 아이스크림을 만드는 제조 기계도 젤라또 기계를 사용하며, 수제 아이스크림과 젤라또 두 제품 모두 만들어지는 과정에서 자연스럽게 부여되는 오버런은 약 25~40%이다.

As for the handmade ice cream made at the store, the method of making and the ingredients used are the same as gelato, but it is often sold at a lower serving temperature or mixed with a lot of toppings. A gelato machine is often used as a handmade ice cream machine, and both handmade ice cream and gelato have an overrun of about 20~40% that is naturally given in the process of making.

공장에서 대량 생산되는 아이스크림의 경우 생산 시스템이 매장의 시스템과 완전히 다르고 사용하는 기계도 다르다. 또한 작업자가 오버런 비율을 조절할 수 있는데, 제품마다 다르긴 하지만 비율에 따라 40~100% 이상으로 부피가 커진다. 대량 생산 아이스크림이 판매되는 온도는 -18°C 이하인데도 질감이 부드러운 이유는 바로 오버런이 많이 들어갔기 때문이다.

The ice cream mass-produced in factories has a production system that is completely different from that of stores, and the machines used are also different. Also, the operator can adjust the overrun rate, which varies from product, but the volume increases more than 40~100% depending on the rate. Even though the temperature at which mass-produced ice cream is sold is below -18°C, the reason why the texture is soft is because it has a lot of overrun.

2 Water, Air (Overrun)

물, 공기, 오버런

젤라또는 수분과 고형분, 그리고 공기의 결합체이다.

수분은 젤라또의 질감을 부드럽게 만들어주고, 얼음 입자를 크게 형성해 거칠게 만들거나, 입 안에서 녹는 속도 등을 결정하는 가장 원초적인 재료이다. 젤라또는 영하의 온도에서 서빙되는 콜드 디저트로 얼어 있어야 하지만 그렇다고 너무 얼어 있으면 안되는, 즉 적절한 수분 함량을 유지하며 고형분과 균형을 맞춰 만들어진다.

젤라또에서 수분이라고 하면 우리가 마시는 물 그 자체를 말하는 것이 아닌, 사용하는 재료들에 포함된 수분의 함량을 이야기한다. 그렇기에 젤라또에 사용되는 재료들의 수분 함량을 정확히 알아야 균형 잡힌 레시피를 작성하고 이에 맞는 젤라또를 제조할 수 있다.

공기가 없는 젤라또는 단단한 얼음 덩어리이다. 적절한 비율의 공기 함량은 젤라또에 부드러운 질감을 주고 입안에서 차갑게 느껴지지 않게 한다.

오버런Overrun은 젤라또의 제조 과정에서 공기가 섞여 부피가 늘어나는 비율을 나타내는 용어로, 오버런의 비율만으로도 젤라또의 맛과 질감이 달라지는 중요한 요소이다. 젤라또의 이상적인 오버런 비율은 25~40%이다.

오버런 계산 방법은 총 중량에서 공기의 부피를 빼고, 그 값을 기존의 중량에 나누어 백분율로 계산한다.

Gelato is a combination of water, solids, and air.

Water is the most essential element that softens the texture of gelato, forms large ice particles to make it rough, or determines the speed it melts in your mouth. Gelato is a cold dessert served at sub-zero temperatures that should be frozen, but not too much, meaning that it maintains adequate water content, balanced with solids.

When we talk about water in gelato, it's not the water we drink but the amount of water contained in the ingredients used. Therefore, if you know precisely the water content of the ingredients used in gelato, you can write a balanced recipe and make gelato suitable for it.

A gelato without air is a solid block of ice. The right proportion of air content gives gelato a soft texture and keeps it from feeling cold in the mouth.

Overrun is a term that indicates the rate at which air is mixed and the volume increases during the gelato-making process, which, just by the overrun ratio, is a critical factor that changes the taste and the texture of gelato. The ideal overrun rate for gelato is 25~40%.

The overrun calculation method is calculated as a percentage by subtracting the volume of the air from the total weight and dividing the value by the existing weight.

액상 믹스 = 150g, 완성된 젤라또 무게 120g일 경우의 오버런 계산

오버런 (%) = ((액상 믹스 무게 – 완성된 젤라또 무게) ÷ 완성된 젤라또 무게) × 100

오버런 (%) = ((150 – 120) ÷ 120) × 100 = 25%

즉, 위 젤라또의 오버런은 25%가 된다.

Liquid mixture = 150 g,
Calculation of overrun when the completed gelato weighs 120 g

Overrun (%) = [(Weight of liquid mixture – Weight of completed gelato) ÷ Weight of completed gelato] × 100

Overrun (%) = [(150 – 120) ÷ 120] = 25%

Therefore, the overrun of the above gelato is 25%.

3 10 Steps to Making Gelato

젤라또 제조 10단계

젤라또를 제조하기 위해서는 10단계의 과정이 필요하다. 젤라떼리아 운영 방향성에 따라 몇 가지 단계가 제외되는 경우도 있지만, 가장 첫 단계인 재료의 선택부터 마지막 단계인 소비자에게 판매되기까지의 과정을 살펴보자.

It takes ten steps to make gelato. A few steps may be excluded depending on the direction of operating a gelateria. First, let's look at the process from the first step of selecting ingredients to the final step of selling to customers.

Step 1 Selecting the Ingredients

재료의 선택

젤라또를 만드는 가장 첫 번째 단계는 '재료의 선택'이다.

재료는 젤라띠에레가 선택하는 재료의 품질, 신선도, 젤라떼리아 운영 콘셉트에 따라 달라지며, 그 선택으로 최종 젤라또의 맛에도 영향을 미친다.

한국은 아직 이탈리아에 비해 젤라떼리아가 많지 않기 때문에(이탈리아 약 38,000개, 한국 약 200개), 젤라떼리아 콘셉트를 세분화하기가 이를 수 있지만, 시장성이 확대되면 다양한 콘셉트의 젤라떼리아가 생길 것이다. 예를 들어 유기농 재료 사용을 메인 콘셉트로 잡으면 젤라떼리아에서 사용되는 재료가 모두 유기농 제품으로 선택될 것이고(한국에서 현재는 일부 재료만 유기농으로 사용하는 매장 존재), 비건으로 콘셉트를 잡으면 동물성 재료가 아닌 식물성 재료만으로 젤라또를 만드는 젤라떼리아가 보편화될 것이다.

어떤 콘셉트의 젤라떼리아든 완성된 젤라또의 맛과 품질을 높이기 위해서는 신선하고 품질 좋은 재료를 선택해야 하며, 그러기 위해서는 재료 공급 업체도 잘 선정해야 한다. 원재료를 사용할 경우 공급 업체의 위생 관리가 철저한지, 페이스트 등 가공 제품을 사용하는 경우 원하는 시기에 공급이 제대로 이루어지는지를 잘 확인해야 한다.

재료의 선택에 따른 재료 관리도 상당히 중요하다. 공급받은 재료들은 제품마다의 적정 보관 온도를 잘 확인하여 그에 맞게 보관하고, 신선 제품들은 소비 기간에 꼭 유의해야 한다.

The first step in making gelato is 'selecting the ingredients.'

The ingredients depend on the quality and freshness of the ingredients and the gelateria operating concept selected by the gelatiere, and the choice affects the taste of the final gelato.

Compared to Italy, there are fewer gelaterias in Korea yet, so it may be too early to subdivide the concept of a gelateria (about 38,000 in Italy and about 200 in Korea). However, as marketability expands, various concepts of gelaterias will emerge. For example, suppose the main concept is to use organic ingredients. In that case, all ingredients used in the gelateria will be organic products (in Korea, some shops currently use only a few ingredients as organic). If the concept is to go vegan, gelaterias, which make gelato with plant-based ingredients rather than animal-based ingredients, will become common.

Regardless of the concepts of gelateria, you should select fresh and high-quality ingredients to enhance the taste and quality of the finished gelato. It is essential to check whether the supplier's hygiene control is thorough when using raw ingredients and whether they properly supply at the desired time when using processed products.

Managing ingredients according to the selection of the ingredients is also quite important. The supplied ingredients must be thoroughly checked for proper storage temperature for each product and stored accordingly, and fresh ingredients must be careful about the consumption period.

Step 2 Writing the Base Recipe

베이스 레시피 작성

쇼케이스에 젤라또가 진열되기 전까지 두 번의 레시피 작업이 필요하다. 바로 그 첫 번째가 젤라또를 만들기 위해 필요한 베이스 레시피 작업이다. 쉽게 말해 빵을 만들 때 반죽이 필요하듯 젤라또의 기본 반죽을 베이스라 부른다.

젤라또를 만드는 제조방법론이 두 가지 방향성으로 나뉘는데, 베이스를 기반으로 젤라또를 만드는 경우 2단계 과정이 중요하다.

주재료를 받아주는 베이스는 한국에서 화이트 베이스, 옐로우 베이스 두 가지로 나뉜다. 젤라띠에레가 선택한 재료들의 다양한 성분이 서로 조화롭게 이루어지게 하기 위한 작업으로, 베이스 레시피가 탄탄하게 균형이 잡혀 있어야 주재료가 첨가되었을 때 균형을 맞추기가 쉽다.

젤라띠에레의 성향과 젤라떼리아 환경에 따라 레시피는 달라지기에 레시피의 절대값이란 없지만, 쇼케이스 내에서 크고 작은 문제점이 발생하지 않는 젤라또를 진열하기 위한 고형분들의 수치 가이드는 존재한다.

Two types of recipes are required before gelato is displayed in the showcase. The very first is to write the base recipe needed to make gelato. In other words, just as the dough is needed to make bread, the basic dough for gelato is called the base.

The manufacturing methodology for making gelato is divided into two directions, and when making gelato based on the base, this 2nd step is important.

In Korea, the base that takes the main ingredient is divided into two types: white and yellow. It's a task to ensure that the various components of the ingredients selected by the gelatiere are harmonized with each other, and it's easy to balance when the main ingredient is added when the base recipe is firmly balanced.

There is no preset value to the recipe, as the recipe varies depending on the gelatiere's inclination and the gelateria environment. However, there is a numerical guide for solids for displaying gelato that does not cause great or small problems in the showcase.

안정적인 젤라또를 완성하기 위한 고형분 수치 가이드

Guide to solids content to make stable gelato

구분 Type	당 Sugar	지방 Fat	무지유 고형분 Non-fat milk solids	기타 고형분 Other solids	총 고형분 Total solids
화이트 베이스 White base	14~19%	3~7%	7~12%	0.2~0.5%	32~36%
싱글, 젤라또 Single, gelato	18~24%	7~12%	7~12%	0.2%↗	36~46%

✦ 옐로우 베이스의 경우 다크 베이스처럼 베이스로 사용되기도 하고, 싱글 레시피로 사용되는 경우도 있어 '완성된 젤라또(싱글, 젤라또)의 고형분 수치'로 본다.

✦ The yellow base is sometimes used like the dark base or as a single recipe. So, it's regarded as the 'solids content of completed gelato (Single, gelato).'

위의 가이드에 따라 베이스 레시피를 작성하면 '맛있다, 맛없다'의 의미가 아닌, 쇼케이스에 진열하기에 이상적인 젤라또의 컨디션이 된다. 젤라또는 과학과 수학이 아닌, 같은 곡을 연주해도 지휘자나 연주자에 따라 곡이 다르게 해석되는 음악과 같다.

Creating a base recipe according to the guide above will be an ideal condition for gelato to display in a showcase, not in the sense of 'delicious or not.' Gelato is not a science or mathematics but akin to music that interprets the same melody differently depending on the conductor or performer.

Step 3 Heating (Pasteurizing)

가열(살균)

베이스 레시피 작성이 끝나면 액상과 분말 재료를 혼합하여 가열(살균)하는 과정이 필요하다. 가열 과정을 통해 레시피상의 분말 재료들이 서로 뭉치지 않게 잘 용해되고, 성질이 서로 다른 성분들이 잘 융합되며 지방도 균질화된다. 위생적인 측면에서는 작업 과정 중에 생겨날 수 있는 몸에 해로운 미생물과 박테리아를 한 번 더 감소시켜주는 역할을 한다.

베이스 가열에는 화이트, 옐로우, 다크 베이스 세 가지의 살균 온도가 있는데, 한국에서는 화이트 베이스와 옐로우 베이스가 주로 사용된다. 다크 베이스의 경우 초콜릿 베이스라고도 불리는데, 한국에서는 다크 베이스로 만들 수 있는 젤라또가 한정적이어서(어떤 주재료를 첨가해도 기본 맛은 초콜릿 맛이다.) 베이스로 제조하는 젤라떼리아가 많지 않다.

① 고온 살균 (화이트 베이스)
젤라떼리아에서 가장 많이 사용되는 화이트 베이스는 살균기, 복합가열제조기 또는 냄비에 우유를 먼저 넣고, 35~40°C로 가열이 되면 분말류(당, 탈지분유, 안정제)를 넣은 후 45°C에 생크림을 넣는다. 그 다음 85°C까지 가열시킨 후 그 온도를 급속으로 4°C로 낮추는 냉각을 한다.

② 저온 살균 (옐로우 베이스)
노른자가 들어가는 옐로우 베이스는 살균기, 복합가열제조기 또는 냄비에 우유와 노른자를 먼저 넣고, 35~40°C로 가열이 되면 분말류(당, 탈지분유, 안정제)를 넣은 후 45°C에 생크림을 넣는다. 그 다음 65°C까지 가열시킨 후 온도를 유지하며 30분 동안 휴지기를 가지고 급속으로 4°C로 낮추는 냉각을 한다.

③ 초고온 살균 (다크 베이스, 초콜릿 싱글 레시피)
초콜릿이 들어가는 다크 베이스는 살균기, 복합가열제조기 또는 냄비에 우유를 먼저 넣고, 35~40°C로 가열이 되면 분말류(당, 탈지분유, 카카오파우더, 안정제)를 넣은 후 45°C에 생크림을 넣는다. 그 다음 90°C까지 가열시킨 후 85°C로 온도가 내려갈 때 초콜릿을 넣고 급속으로 4°C로 낮추는 냉각을 한다. 초콜릿을 처음부터 넣지 않고 가열 후 냉각 과정에 넣는 이유는 초콜릿의 풍미를 해치지 않게 하기 위함이다.

젤라띠에레가 만들고자 하는 베이스에 따라 가열 온도가 다르지만, 공통적으로 분말류는 베이스 완성 시 뭉침 현상을 방지하기 위해 35~40°C의 따뜻한 온도에 넣는다. 또한 생크림은 풍미를 얼마만큼 부여하고 싶은지에 따라 우유가 들어가는 온도 4°C, 가열되는 온도 45°C, 가열 후 냉각되는 온도 65°C에 첨가한다. (생크림이 들어가는 온도에 따라 최종 베이스의 풍미가 달라진다.)

After writing the base recipe, you need to combine and heat (pasteurize) the liquid and powder ingredients. It lets the powder ingredients in the recipe dissolve well through the heating process so they won't get lumped, the ingredients with different properties fuse well, and the fat is homogenized. In terms of hygiene, it helps to reduce once again harmful microorganisms and bacteria that can occur during the working process.

There are three pasteurizing temperatures for heating the base; white, yellow, and dark. White and yellow bases are commonly used in Korea. As for the dark base, also called the chocolate base, limited gelato can be made with dark base in Korea, so few gelaterias are using the base. (Whichever main ingredient you add, the base flavor is chocolate.)

① High-temperature pasteurization (White base)
To make the white base most commonly used in gelaterias, first put milk in a pasteurizer, multi-heater, or pot. When it reaches 35~40°C, add powders (sugar, skim milk powder, stabilizer) and then add cream at 45°C. Heat to 85°C and rapidly cool to 4°C.

② Low-temperature pasteurization (Yellow base)
For the yellow base that contains egg yolks, put milk and egg yolks first in a pasteurizer, multi-heater, or pot. When it reaches 35~40°C, add powders (sugar, skim milk powder, stabilizer) and then add cream at 45°C. Heat to 65°C, let it rest while maintaining the temperature, and rapidly cool to 4°C.

③ Ultra-high temperature pasteurization (Dark base and Chocolate single recipe)
To make the dark base with chocolate, first put milk in a pasteurizer, multi-heater, or pot. When it reaches 35~40°C, add powders (sugar, skim milk powder, cacao powder, stabilizer) and then add cream at 45°C. Heat to 90°C, and when the temperature drops to 85°C, add chocolate to mix and rapidly cool to 4°C. The reason why chocolate is not added from the beginning but in the cooling process is not to degrade the flavor of the chocolate.

The heating temperature differs depending on the base the gelatiere wants to make, but powders are commonly added at a warm temperature of 35~40°C to prevent lumps. As for the cream, depending on how much flavor of the fresh cream you want to convey, you can add it at 4°C when adding the milk, at 45°C when heating, or at 65°C when it's cooling after heating. (The flavor of the base varies depending on the temperature at which the cream is added.)

Step 4 Aging

숙성

젤라띠에레마다 숙성 과정의 선택 기준이 다르지만, 와인과 치즈 또는 우리의 김치와 장류는 숙성시켜 먹어야 풍미가 더 깊어지는 것처럼, 베이스 역시 가열 후 냉각한 베이스를 4°C로 계속 유지하여 숙성시켜야 보다 더 풍미가 깊고 질감이 안정적인 젤라또로 완성된다. 또한 베이스에 들어가는 당, 유지방, 단백질 등이 완전히 용해되며 수분과 서로 잘 융합되는 시간을 가지며 고형분 간의 밀도가 높아져 안정제가 완전한 작용을 하도록 한다.

모든 고형분들이 따로 놀지 않고 하나로 어우러져야 큰 얼음 결정체가 아닌 아주 미세하고 부드러운 얼음 결정체로 형성되고, 자연적인 공기의 혼합으로 젤라또를 먹었을 때 입 안에 들어오는 느낌이 너무 차갑지 않으면서 부드럽게 유지된다.

숙성 시간은 최소 1~2시간이 필요하며, 숙성 베이스의 최대 보관 시간은 72시간이다. 가장 이상적인 숙성 시간은 10~12시간(하룻밤)으로, 일반적으로 영업 마감 후 베이스를 만들어 숙성시키고, 다음날 오전에 숙성된 베이스로 젤라또 제조를 한다. 베이스 숙성이 젤라또 제조 과정의 필수 과정이라고 할 수는 없지만 좋은 품질의 젤라또를 만들기 위해서는 충분히 필요한 시간이다.

Each gelatiere has different criteria for selecting the aging process. Nonetheless, just as wine, cheeses, kimchi, and sauces must be aged to deepen their flavor, you can achieve deeper flavored gelato with a stable texture by storing the cooled base at 4°C. Additionally, sugar, milk fat, protein, and other components added to the base can completely dissolve and take its time to infuse well with moisture. Then the density between solids components increases, allowing the stabilizer to function fully.

Very fine and soft ice crystals are formed when all the solids are combined without separating, not large ice crystals. And the natural aeration keeps the gelato in your mouth soft without being too cold.

A minimum of 1~2 hour is required for aging, and the maximum storage period for aging the base is 72 hours. The ideal aging time is 10~12 hours (overnight), and generally, the base is made and aged after closing the shop, and make the gelato with the aged base the following day. Aging the base is not an essential process in making gelato, but it is a necessary time to make a good quality gelato.

Step 5 Deciding on The Flavor

맛의 결정

가열(살균) - 숙성 과정을 거쳐 베이스가 완성된 후 젤라또로 만들기 위해 주재료를 결정하는 단계이다. 1단계 재료 선택의 연장선이라 할 수 있는 이 단계에서, 젤라띠에레들의 재료 사용에 대한 성향과 젤라떼리아의 콘셉트를 결정할 수 있다. 예를 들어, 바닐라 젤라또를 만들기 위해 A 젤라띠에레는 바닐라빈(원재료)을 인퓨징해서 만들고, B 젤라띠에레는 바닐라 페이스트(가공 재료)를 사용하고, C 젤라띠에레는 이 두 가지를 섞어서 만든다. 즉, 스스로가 원하는 최종 젤라또 맛에 따라 결정하고 자신만의 젤라또를 표현하는 단계라고 볼 수 있다.

주재료를 받아주는 베이스가 화이트 베이스인지 옐로우 베이스인지에 따라서도 같은 재료가 쇼케이스에 진열되는 최종 젤라또의 맛은 완전히 달라진다.

예 1. 화이트 베이스 + 바닐라 = 깔끔한 맛
예 2. 옐로우 베이스 + 바닐라 = 풍부한 맛

It is the step to determine the main ingredients to make gelato after making the base through heating (pasteurizing) and the aging process. This step is an extension of the first step of selecting the ingredients, where the gelatieres can determine the ingredients and the concept of gelaterias by their inclinations. For example, to make vanilla gelato, an A gelatiere would infuse vanilla bean (raw ingredient), a B gelatiere would use vanilla paste (processed ingredient), and a C gelatiere would use a mixture of the two. Hence, this can be a step to deciding your desired final gelato taste and expressing your own.

Depending on whether the main ingredient will be mixed with a white or yellow base, the taste of the final gelato displayed in the showcase will be completely different, even with the same ingredient.

E.g. 1. White base + Vanilla = Fresh taste
E.g. 2. Yellow base + Vanilla = Rich taste

Step 6 Writing the Gelato Recipe

젤라또 레시피 작성

2단계에서 언급한 두 번의 레시피 중 두 번째 레시피 작성의 단계이다. 2단계에서 베이스의 고형분을 균형 있게 맞춘 상태인데 여기에 다시 주재료를 첨가하면 주재료 성질에 따라 균형이 무너진다. 따라서 수치가 변화된 고형분을 쇼케이스에 진열하기 이상적인 고형분 수치로 다시 보완하는 레시피 작성이 필요하다.

베이스에 주재료를 첨가해서 젤라또를 만드는 젤라띠에레가 아닌, 주재료의 성질에 따라 각각의 젤라또 믹스를 만드는 경우(싱글 레시피) 2단계 과정이 아닌 6단계의 레시피 작성 과정을 거쳐 가열(살균), 숙성의 단계로 간다.

It is the step to create the second recipe among the two recipes I mentioned in step 2. In the first step, the solids content of the base is balanced, but when the main ingredient is added, the balance can break apart depending on the nature of the ingredient. Therefore, you should write the recipe that supplements the changed solids content with an ideal solids content value for putting in the showcase.

Suppose you make each gelato mix according to the main ingredient's properties (single recipe) rather than adding the main ingredient to the base. If so, you can skip to step 6 instead of applying the two-step process, followed by heating (pasteurizing) and aging.

안정적인 젤라또를 완성하기 위한 고형분 수치 가이드

Guide to solids content to make stable gelato

구분 Type	당 Sugar	지방 Fat	무지유 고형분 Non-fat milk solids	기타 고형분 Other solids	총 고형분 Total solids
화이트 베이스 White base	14~19%	3~7%	7~12%	0.2~0.5%	32~36%
싱글, 젤라또 Single, gelato	18~24%	7~12%	7~12%	0.2%↗	36~46%

✦ 위 표에서 '싱글, 젤라또'라 함은 싱글 레시피 젤라또, 베이스로 만든 젤라또 레시피를 말한다. (즉, 쇼케이스에 들어가는 '완성된' 젤라또의 고형분 수치이다.)

✦ In the above table, 'single, gelato' refers to a single recipe gelato or a gelato recipe made with a base. (In other words, it is the solids content of 'completed' gelato that goes into the showcase.)

Step 7 Making

제조

액체 상태의 젤라또 믹스를 냉각 교반 과정을 통해 젤라또로 제조하는 단계이다.

4°C 숙성을 거치고 주재료가 첨가된 젤라또 믹스 또는 싱글 젤라또 믹스를 제조기에 넣고 기계를 작동하면, 제조기 실린더의 온도가 -30°C 이하로 급격히 내려가며 냉각 교반이 되는 동시에 오버런이 25~40% 진행된다. 소요 시간은 젤라또 믹스의 양과 고형분 함량, 제조기 성능에 따라 차이가 있지만 젤라또 완성까지 7~10분 정도 소요된다. 젤라또 완성 시 추출 온도는 -7~-9°C이고 차갑게 냉각된 바트나 카라피나에 원하는 모양으로 추출한다.

Now, you will turn the liquid gelato mix into gelato through a cooling and churning process.

Pour the gelato mix or single gelato mix that has been aged at 4°C with the main ingredients into the maker and operate the machine. Then the temperature of the freezer's cylinder will rapidly drop below -30°C, and cooling and churning will start as well as the overrun by 25~40% simultaneously. The time required varies depending on the amount of gelato mix, the solids content, and the performance of the freezer, but it takes about 7~10 minutes to finish. When finished, the extraction temperature is -7~-9°C, and you can extract it in the desired shape in a container or a carapina that has been kept frozen.

Step 8 Blast Freezing

급속 냉동

바트나 카라피나에 담긴 젤라또는 보기와는 다르게 미완성의 상태이다. 젤라또는 수분과 고형분, 공기의 결합체로 추출된 젤라또 내부에는 얼지 않은 수분이 존재한다. 바로 추출한 젤라또를 쇼케이스에 진열하면 얼지 않은 수분들이 높은 온도에서 천천히 얼어 얼음 입자가 크게 만들어지고 수분과 공기가 증발하는 현상이 발생한다. 또한 쇼케이스 내부에서 젤라또가 녹으며 다시 얼기 때문에 무너지는 볼륨과 서걱거리는(아이시한) 질감이 형성된다. 이러한 치명적인 부작용들을 방지하기 위해 젤라또에서는 급속 냉동의 단계가 중요하다.

바트나 카라피나에 추출한 젤라또를 급속 냉동고에 넣고 5~10분 정도 급속으로 냉동시키면 젤라또 표면의 수분과 공기가 증발하는 것이 방지되는 보호막이 한 겹 형성된다.

1시간 이상 급속 냉동을 할 경우 젤라또 표면뿐 아니라 중심부까지 급속으로 냉각되므로, 바로 진열해서 판매하는 젤라또가 아닌 -18°C 이하 냉동고에 보관 후 판매하는 젤라또에 적합하다.

Contrary to what it looks like, the gelato in the container or carapina is yet to be finished. Gelato has been extracted in the combined form of moisture, solids, and air, and unfrozen moisture exists in the center of gelato. If you put the freshly extracted gelato in a showcase, unfrozen moisture freezes slowly at a high temperature, makes large ice particles, and evaporates moisture and air. Also, as the gelato melts and freezes again in the showcase, volume collapses and forms a coarse (icy) texture. Blast freezing is an essential step in making gelato to prevent these fatal side effects.

When gelato is extracted into a container or carapina and placed in a blast freezer to freeze rapidly for 5~10 minutes, a protective film is formed on the surface of the gelato to prevent moisture and air from evaporating.

If you blast freeze for more than one hour, not only the surface of the gelato but also the center will freeze rapidly. So, unless you are planning to display and sell immediately, it is suitable for gelato to sell after storing frozen under -18°C.

Step 9 Storing Frozen

냉동 보관

젤라또의 냉동 보관은 오픈형 쇼케이스를 사용하는 젤라떼리아에서 필수적인 단계이다. 마감 후 젤라또를 쇼케이스에서 -18°C 이하 또는 매장의 쇼케이스보다 1~2°C 낮은 냉동고에 보관하는 과정이다. 젤라또를 재고로 적재하고 판매하는 젤라떼리아의 경우 냉동 보관 관리를 철저하게 해주어야 질감과 맛의 변화가 없는 젤라또를 판매할 수 있다.

-18°C 이하의 냉동고에서 보관하던 젤라또를 -12°C의 쇼케이스에 옮겨 바로 판매할 수는 없다. 젤라또 온도 1°C의 변화를 주는 데 40분~1시간이 소요되므로 이와 같은 경우 4~6시간 동안 쇼케이스에 진열해놔야 부드러운 질감의 젤라또를 판매할 수 있다. 따라서 효율적으로 젤라또 온도를 관리하는 젤라떼리아에서는 냉동 보관이 가능한 냉동고를 1대만 사용하지 않고 -18°C 이하 냉동고, 매장의 쇼케이스보다 1~2°C 정도 낮게 온도 설정된 냉동고를 함께 사용한다. 온도가 다른 냉동고를 2대로 운용할 경우 젤라또에 가장 치명적인 열 충격을 강하게 주지 않고 천천히 온도 변화를 주며 관리와 판매를 할 수 있다.

Freezing gelato is an essential step in gelaterias using open showcases. It is a process of storing gelato in a freezer under -18°C or in a freezer that is 1~2°C lower than the showcase after closing the shop. If you keep stocks of gelato to sell, frozen storage management must be thoroughly managed to sell the gelato without change in texture and taste.

You cannot sell the gelato stored under -18°C immediately after moving to a showcase at -12°C. It takes 40~60 minutes to lower the temperature of gelato by 1°C, so in this case, you need to put it in the showcase for 4~6 hours to sell gelato with a soft texture. Hence, gelaterias that efficiently manage the temperature of gelato do not use only one unit of the freezer but use one freezer at -18°C or lower and one at a temperature 1~2°C lower than the store showcase. By operating two freezers with different temperatures, you can manage and sell gelato by slowly changing the temperature without giving the fatal thermal shock.

Step 10 Displaying and Selling

진열 및 판매

한 단계 한 단계를 거쳐 완성된 젤라또를 젤라띠에레가 원하는 방법으로 진열하고 소비자에게 판매하는 단계이다. -10~-14°C로 판매되는 젤라또는 두 가지 방법으로 진열 및 판매가 된다. 소비자의 눈으로 젤라또의 형태와 색감을 보고 맛을 선택할 수 있는 오픈형 쇼케이스, 젤라또를 보지 못하고 맛 이름만 보고 맛을 선택할 수 있는 밀폐형 쇼케이스(포제띠)가 있다.

오픈형 쇼케이스는 눈에 보여지는 것이 중요한 만큼 바트가 항상 청결하게 정돈되어야 하고 외부 공기의 접촉과 쇼케이스 온도 변화에 항상 신경 써야 한다. 쇼케이스에 12가지의 젤라또가 진열되어 있고 한 가지 맛을 서빙하더라도 나머지 11가지 맛들까지 외부 공기와 온도에 접촉이 된다. 쇼케이스 문을 열고 닫는 반복되는 과정 속에서 젤라또의 품질은 점차 떨어져 판매 기간이 짧다.

반면에, 밀폐형 쇼케이스(포제띠)는 시각적인 요소는 없지만 젤라또 맛마다 뚜껑이 덮어져 진열되기에 일정한 온도를 유지할 수 있고, 외부 공기의 접촉률이 낮아 품질 저하가 늦고 오픈형 쇼케이스보다 판매 기간이 길다.

오픈형 쇼케이스와 밀폐형 쇼케이스는 젤라떼리아의 얼굴이라 할 만큼 중요한 부분이 위생이므로, 위생 관리도 철저하고 인테리어적으로도 잘 어우러지게 배치되어야 한다.

젤라또 판매는 콘 또는 컵이 사용되는데 한국에서는 컵을 주로 선호하는 편이다. 젤라또 외 콜드 디저트를 함께 판매하는 경우에는 각각의 디저트에 어울리는 냉동 온도에 진열 및 판매를 해야 한다.

-10~-14°C : 젤라또, 소르베또, 셔벗

-18°C 이하 : 젤라띠니, 보틀 젤라또, 젤라또 케이크, 스틱 젤라또

-4~-6°C : 그라니따

It is the step to display the gelato, which has been completed step by step in a way the gelatiere wants and sells to customers. Gelato can be displayed and sold in two ways at -10~-14°C. One is in an open showcase where customers can see the form and color of gelato and select a flavor. The other is a closed showcase (Pozzetti) where customers can choose the flavor by name without seeing the gelato.

As the visual aspect is vital for an open showcase, the container must always be kept clean, and always pay attention to the contact of outside air and the temperature changes of the showcase. If there are 12 flavors of gelato displayed in the showcase, and even if one flavor is served, the remaining 11 flavors will also come in contact with the outside air and temperature. In the repeated process of opening and closing the showcase, the quality of gelato gradually deteriorates, resulting in a short selling period. On the other hand, the closed showcase (Pozzetti) has no visual elements, but it can maintain a constant temperature because each flavor is covered and displayed.
So, the quality deterioration is slow due to less contact with the outside air, and the selling period is more extended than that of an open-type showcase.

Open and closed (Pozzetti) showcases are important enough to represent the gelateria; therefore, hygiene management should be thorough and in tune with the interior.

Cones and cups are used when selling gelato; usually, cups are preferred in Korea. If other cold desserts are sold along with gelato, those must be displayed and sold at a freezing temperature suitable for each dessert.

-10~-14°C : Gelato, sorbetto, sherbet

Under -18°C : Gelatini, bottle gelato, gelato cake, gelato stick

-4~-6°C : Granita

How to Shape and Fill Gelato in a Showcase Container

쇼케이스용 바트에 젤라또를 모양 내어 담는 방법

충전물이나 토핑이 있는 젤라또 담기 Putting gelato with fillings or toppings

① 추출되고 있는 젤라또를 바트 1/3 높이까지 받는다.

② 충전물을 넣고 스패출러로 섞어준다.

③ 동일한 과정으로 ①과 ②를 두 번 더 반복한다.

④ 마지막 윗면의 젤라또를 스패출러로 정리한다.

⑤ 토핑이 있는 경우 윗면에 뿌려준다.

① Receive the gelato being extracted up to 1/3 of the container.

② Add filling and mix using a spatula.

③ Repeat ① and ② two more times the same way.

④ Organize the top of the final extraction with a spatula.

⑤ If you are using toppings, sprinkle them on top.

물결 모양으로 젤라또 담기 Putting gelato in a wave pattern

① 추출되고 있는 젤라또를 기계에서 나오는 모양 그대로 살려주며 스패출러로 끊어가며 바트에 담는다.

② 바트 1단에 왼쪽 – 오른쪽 – 왼쪽 – 오른쪽 순서로 젤라또를 담는다.

③ 바트 2단도 동일한 방법으로 젤라또를 담는다.

④ 마지막 단은 가운데 한 줄로 젤라또를 담거나 1단, 2단과 동일한 방법으로 담는다.

⑤ 마지막에 조금씩 나오는 젤라또는 모양을 살리기 어려우므로 서빙 스쿱이 꽂힐 자리에 담는다.

① Take the extracted gelato while preserving its natural pattern as it comes out of the freezer, cut it with a spatula, and put it in a container.

② Place the gelato in the order of left - right - left - right on the first layer.

③ Repeat the same on the second layer.

④ Finish the last layer in one row in the middle; or the same way as the previous layers.

⑤ Since preserving the shape of the last bits that extract little by little is difficult, place it where you will stick the serving scoop.

둥근 모양으로 젤라또 담기 Putting gelato in a round shape

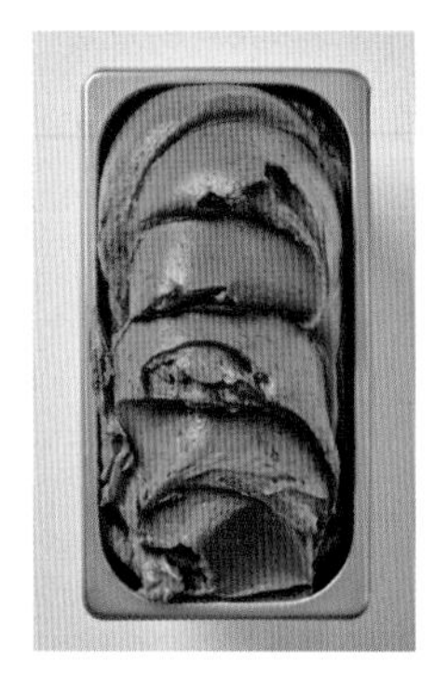

① 추출되고 있는 젤라또를 바트에 담으면서 스패출러로 넓게 펴준다.

② 동일한 방법으로 반복하며 끝까지 담는다.

③ 스패출러로 모양을 전체적으로 둥글게 잡아준다.

④ 깨끗한 젤라또 서빙 스쿱을 이용해 힘을 살짝 주며 모양을 잡아 마무리한다.

① Spread widely with a spatula while receiving the gelato being extracted.

② Repeat the same way until the end.

③ Shape the entire gelato round using a spatula.

④ Use a clean gelato serving scoop to give it a bit of force and shape it to finish.

카라피나(원통형)에 젤라또 담기 Putting gelato in a carapina (cylinder)

① 원통형 바트에 젤라또를 담을 때는 모양을 신경 쓰지 않고 추출되는 젤라또를 차곡차곡 담는다.

② 스패출러로 윗면의 젤라또를 깔끔하게 정리해 마무리한다.

① Put the extracting gelato one after another without worrying about the shape in a cylindrical container.

② Organize the top of the gelato with a spatula to finish.

How to Serve Gelato in Cones and Cups

젤라또를 콘과 컵에 담는 방법

콘과 컵에 젤라또 담기 Serving gelato in cones and cups

① 젤라또 서빙 스쿱으로 젤라또를 넓게 펼치고 모아 풀어준 후 달걀 모양으로 만든다.

② 콘이나 컵에 스쿱을 직선으로 내리면서 모양 낸 젤라또를 담는다.

① Spread the gelato widely with a gelato serving scoop, gather, loosen up, and shape it into an egg shape.

② Pull the shaped gelato down into a cone or cup by lowering the scoop in a straight line.

Cone

Cup

Part 2.

Gelato Making Machines and Tools

젤라또를 만드는 기계와 도구들

1 # Batch Freezer

제조기

젤라또를 만드는 젤라또 제조기는 3가지로 나눌 수 있다. ① 완성된 믹스(젤라또 재료가 모두 혼합된 것)를 냉각 교반 제조만 하는 '냉각 제조기', ② 가열과 냉각 기능이 있어 베이스 가열과 젤라또 냉각 교반이 가능한 '가열 복합 제조기', ③ 젤라또 제조와 완성된 젤라또 진열 및 보관이 가능한 '쇼케이스 복합 제조기'가 있다. 제조기는 젤라떼리아의 운영 형태나 규모, 인테리어에 따라 선택할 수 있다.

또한 제조기의 용량에 따라 한 번에 한 바트만 생산하거나 두 바트 이상 동일한 맛을 연속 생산할 수 있어, 효율적인 노동 시간과 에너지 대비 생산성을 가져가기 위해 기계 용량 선택도 잘 해야 한다.

✦ 기계 세척이 용이한 물 호스 결합 기계를 선택할 경우 온수 연결을 추천한다.

The batch freezer that makes gelato can be classified into three types: ① A cooling freezer that only cools and churns the completed mixture (a mixture of all gelato ingredients), ② A heating combined freezer with a heating and cooling function that can heat the base and cool-churning the gelato, and ③ A showcase integrated freezer capable of producing gelato and displaying and storing the completed gelato. You can select the freezer according to the type of operation or the interior layout of the gelateria.

Also, depending on the capacity of the gelato freezer, it can produce one batch at a time or the same flavor for more than two batches continuously, so it is necessary to select the machine capacity wisely to control labor time efficiently and energy versus productivity.

✦ I recommend connecting hot water when choosing a water hose coupling machine that makes cleaning the machine easier.

① 냉각 제조기

냉각 제조기는 고전적인 제조 방식인 '수직형'과 현대적인 제조 방식인 '수평형'으로 나뉜다. 첫 자동 냉각 제조기는 수직 형태로 나무통에 믹스를 넣고 주걱으로 저어가며 젤라또를 만들던 옛날 방식을 그대로 착안하여 만든 제조기이다. 최초의 자동 냉각 제조기로 볼 수 있지만 젤라또가 완성되었음을 자동으로 판단해주는 수평형 제조기가 나온 이후에는 반자동 제조기라고 보는 것이 더 맞겠다. 수평형 제조기와는 달리, 수직형 제조기는 젤라띠에레가 직접 젤라또의 질감을 보며 완성됨을 판단해야 한다. 한 가지 맛이 만들어지는 데 소요되는 시간은 15~25분 정도이다.

수직형 제조기는 젤라또의 완성도를 판단할 수 있는 경험이 부족하면 기계 사용에 익숙해지기까지 시간이 걸리고 젤라또의 완성도도 떨어질 수 있다.

수평형 제조기는 수직형 제조기와 다르게 젤라띠에레가 젤라또의 전체 질감을 눈으로 직접 볼 수는 없지만 경도, 온도, 시간 등을 조절해 자동으로 젤라또의 완성을 알려준다. 젤라띠에레가 직접 완성된 젤라또를 퍼 담아야 하는 수직형 제조기와 달리, 수평형 제조기는 비터의 교반 속도를 높이며 보다 편하게 젤라또를 추출할 수 있다. 한 가지 맛이 만들어지는 데 소요되는 시간은 8~10분 정도이다.

① Batch Freezer

There are two types of cooling freezers; (1) 'vertical type' with a classical production method and (2) 'horizontal type' with a modern production method. The early self-cooling freezer was created based on the old technique of making gelato by putting the mix in a vertical wooden barrel and stirring it with a spatula. You can consider it the first automatic batch freezer, but seeing it as semi-automatic would be more accurate after the horizontal freezer came out. Unlike the horizontal freezer, you have to judge whether gelato is completed by directly looking at the texture of gelato when using the vertical freezer. It takes about 15~25 minutes to make one flavor.

The vertical freezer takes time to get used to using it if you lack the experience to judge the completion of gelato, and the level of completion may be inferior.

Unlike the vertical freezer, you cannot directly see the entire gelato texture in the horizontal freezer. But it automatically informs the completion by adjusting the hardness, temperature, and time.

While gelatiere has to scoop up the finished gelato themselves from the vertical freezers, the horizontal type can extract the gelato more conveniently by increasing the stirring speed of the paddle. It takes about 8~10 minutes to make one flavor.

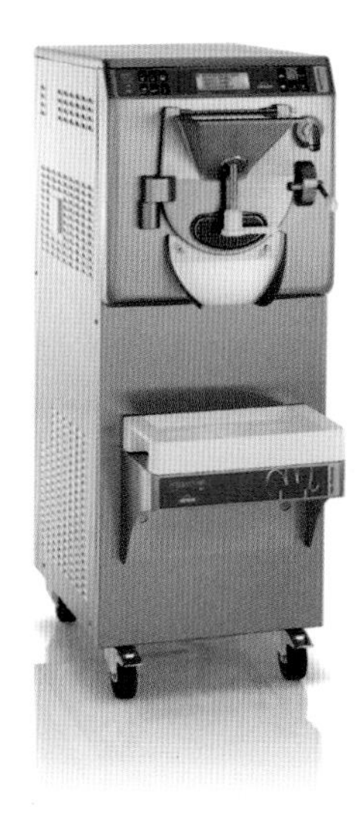

② 가열 복합 제조기

가열 복합 제조기(수평형 제조기)는 베이스를 사용한 레시피 또는 싱글 레시피로 젤라또를 만들 수 있도록 가열 기능과 냉각 기능이 결합된 제조기이다. 젤라또 생산량이 많지 않거나 작업 공간이 협소할 경우 살균기와 제조기를 따로 구비하지 않고 복합제조기 한 대로 젤라떼리아를 운영할 수 있다. 또한 가열 복합 제조기 모델에 따라 요거트, 각종 소스, 초콜릿 템퍼링, 커스터드 크림 등을 만들 수 있는 멀티 기능 제조기도 있다. 매장에서 젤라또만 판매하지 않고 다른 디저트류도 함께 판매한다면 제조기 한 대로 많은 제품들을 만들 수 있어 가격 대비 효율성이 높은 제조기이다.

② Heating Combined Batch Freezer

The heating combined batch freezer(Horizontal freezer) combines heating and cooling functions to make gelato with a recipe using a base or single recipe. If there is not much gelato production or you have a limited working space, you can operate the shop with a single combined freezer without a separate sterilizer and freezer. In addition, there is a multi-function freezer that can make yogurt, various sauces, custard cream, tempered chocolate, etc., depending on the model of the heating combined model. If the store sells not only gelato but also other desserts, it is a highly cost-effective maker because you can make many products with one machine.

③ 쇼케이스 복합 제조기

수직형 제조기와 쇼케이스가 하나로 합쳐진 제조기이다. 밀폐형 쇼케이스처럼 각각 통이 나뉘어 수직으로 삽입된 실린더에 믹스를 넣으면 냉각 교반이 되고, 젤라또가 완성되면 쇼케이스로 따로 옮기는 것이 아니라 그 상태 그대로 보관되며 판매하는 제조기이다. 다른 제조기와 다르게 고객에게 방금 만들어진 신선한 젤라또를 보여줄 수 있고, 액상 믹스에서 젤라또로 만들어지는 변화의 순간을 볼 수 있는 재미도 준다.

③ Showcase Integrated Batch Freezer

It is a gelato freezer in which a vertical machine and a showcase are integrated into one. Like a sealed showcase, each barrel is divided, and you would put the mixture into a vertical cylinder to cool and churn. When the gelato is completed, you don't have to transfer it to a showcase, but it is stored and sold as it is. Unlike other freezers, you can show the customers freshly made gelato, and it is also entertaining to see the moment it changes from a liquid mixture to a gelato.

2 Pasteurizer

살균기

◆ 젤라떼리아에서 사용되는 살균기는 30L 또는 60L 용량이 적당하다. 그 이상의 용량은 보통 대량 생산을 하는 제조 공장에서 사용된다.

◆ A 30-liter or 60-liter capacity pasteurizer is suitable to use in a gelateria. Anything above that is the capacity used in manufacturing factories with high-volume production.

살균기는 젤라또 베이스를 만들거나 싱글 레시피 젤라또 믹스를 대량으로 만들 때 사용하는 기계이다. 당, 유지방, 단백질 등 사용되는 재료 각각의 고형분 특성을 유지하면서 가열을 통해 미생물 활동을 억제시키고, 지방구 입자를 작게 유화시켜 베이스와 믹스의 풍미를 극대화시켜줄 수 있는 기계이다. 90°C의 초고온, 85°C의 고온, 75°C의 중온, 65°C의 저온 살균 등 원하는 가열 온도로 설정이 가능하고, 4°C로 냉각한 후 냉장 숙성 모드로 전환된다. 살균기의 모델에 따라 요거트, 시럽, 치즈 제조도 가능하다.

A pasteurizer is a machine to make gelato bases or large quantities of single-recipe gelato mixes. It's a machine that can maximize the flavor of the base and combine by suppressing microbial activity through heating and emulsifying small fat globule particles while maintaining the solid content characteristics of each ingredient used, such as sugar, milk fat, and protein. It can be set to a desired heating temperature such as an ultra-high temperature of 90°C, high temperature of 85°C, medium temperature of 75°C, pasteurization at 65°C, etc., and after cooling to 4°C, converted to refrigeration aging mode. Depending on the model of your pasteurizer, it also has yogurt, syrup, and cheese-making capabilities.

3 Showcase

쇼케이스

밀폐형 쇼케이스 (포제띠)
closed showcase (Pozzetti)

오픈형 쇼케이스
open showcase

젤라또를 진열하는 쇼케이스는 젤라또가 보이는 오픈형 쇼케이스와 젤라또가 보이지 않는 밀폐형 쇼케이스로 나뉜다.

오픈형 쇼케이스는 고객이 젤라떼리아에 들어왔을 때 눈으로 젤라또를 보는 즐거움을 먼저 느끼게 해주는 장점이 있어 인테리어 요소로도 시각적인 효과를 극대화시켜준다. 반면 젤라또를 서빙할 때 젤라또의 온도 마찰이 잦아 품질이 빨리 저하된다는 단점이 있으며, 판매할 수 있는 기간이 최대 3~5일이다. 오픈형 쇼케이스에는 사각 바트가 들어가며 바트 사이즈에 따라 진열 개수를 조절할 수 있다.

밀폐형 쇼케이스(포제띠)는 젤라또를 눈으로 보는 즐거움이 없는 단점이 있다. 하지만 각각의 젤라또마다 뚜껑이 덮어져 있어 오픈형 쇼케이스에 진열되는 젤라또보다 외부 온도 마찰이 적어 품질 유지 및 관리가 용이하다는 장점이 있다. 판매할 수 있는 기간은 최대 5~7일이며, 밀폐형 쇼케이스는 '카라피나'라고 불리는 원통이 1단 또는 2단으로 들어간다. 2단 밀폐형 쇼케이스를 사용할 경우 보통 아래의 단은 젤라또를 보관하는 단으로 사용되지만, 많은 맛을 진열하고 싶은 매장은 1단과 2단을 다른 맛으로 진열해 판매하기도 한다.

Showcases displaying gelatos are divided into open showcases where gelatos can be seen and closed types where gelatos cannot be seen.

The open showcase has the advantage of allowing customers first to experience the pleasure of seeing the gelatos with their eyes when they enter the gelateria, so it maximizes the visual effect as an interior design element. However, when serving gelato, the temperature friction of gelato occurs frequently and causes the quality to deteriorate quickly; therefore, the selling period is up to 3~5 days. The open showcase holds rectangular-shaped stainless-steel containers, and the number of displays can be adjusted according to the container size.

The closed showcase (Pozzetti) has the disadvantage of not being able to see the gelatos with your eyes. However, since each gelato is covered with a lid, it has the advantage of easy quality maintenance and management due to less temperature friction than the gelatos in the open showcase, so the selling period is up to 5~7 days. The closed showcase can stack a cylinder container called a 'carapina' in one or two tiers. When using a two-tier sealed showcase, the lower level is usually used to store gelato, but shops that want to display many flavors and sell more would arrange different flavors in both tiers.

4 Blast Freezer

급속냉동고

젤라또는 온도 변화에 매우 민감하다. 급속냉동고는 이러한 젤라또를 쇼케이스에 진열하거나 냉동 보관할 때 품질의 변화를 최소한으로 줄이고, 좋은 상태로 유지되도록 도움을 준다. 바트나 카라피나에 젤라또를 추출한 후 바로 판매하는 젤라또의 경우 5~10분간 급속 냉동을 한 후 쇼케이스로 옮기며, 냉동 보관을 위한 젤라또의 경우 1시간 동안 급속 냉동을 한 후 -18°C 이하의 냉동고로 옮긴다.

젤라또 케이크, 스틱 젤라또 등의 콜드 디저트를 만들 때 급속냉동고는 필수적으로 필요하다.

Gelato is very sensitive to temperature changes. A blast freezer helps keep gelato in good condition while minimizing changes in quality when displayed in a showcase or stored frozen. Blast freeze for 5 to 10 minutes and then move it to the showcase to sell the gelato immediately after extracting it into a stainless-steel container or carapina. When storing the frozen gelato, blast freeze for one hour and then move it to a freezer below -18°C.

A blast freezer is a must if you plan to make cold desserts such as gelato cake or stick gelato.

5 Small Tools

기타 소도구

① 저울

젤라또를 제조할 때 중량을 정확하게 지키지 않으면 균형 잡힌 레시피로 만들지 못하므로 젤라띠에레가 원했던 젤라또와 다르게 완성될 수 있다. 사용하는 저울은 1g 단위로 올라가고 30kg까지 여유 있게 계량할 수 있는 전자 저울을 추천한다.

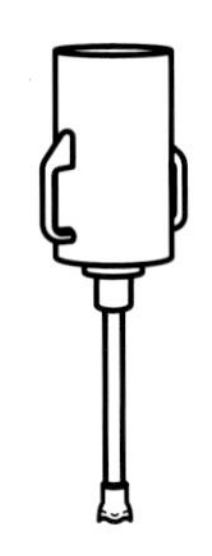

② 핸드블렌더

재료들을 완전히 하나로 혼합해주는 도구로 가정용 핸드블렌더가 아닌 와트수가 높은 업소용 제품을 사용하는 것을 권한다. 핸드블렌더의 성능에 따라 젤라또 믹스의 혼합 결과가 달라지고, 완성된 젤라또의 질감과 맛에도 영향을 미칠 수 있어 매우 중요한 도구이다.

③ 온도계

주로 젤라또 베이스나 젤라또 믹스를 만들 때 온도를 측정하기 위해 사용된다. 인퓨징하는 젤라또를 만들 때에도 온도 체크용으로 필요한 필수 도구이다.

④ 인덕션

젤라또 제조는 불 사용이 많은 작업은 아니므로 가스레인지를 설치할 필요는 없지만 인퓨징하는 젤라또, 부재료(쌀 조리, 시럽, 캐러멜 소스 등)를 사용하는 젤라또, 젤라또 베이스(화이트 베이스, 옐로우 베이스 등), 젤라또 믹스 등을 가열하기 위해 인덕션 하나 정도는 구비해두는 것이 좋다.

⑤ 당도계

소르베또 제조 시 사용하는 과일의 당을 확인하는 용도로 사용하는 중요한 도구이다. 생과일을 사용할 경우 같은 과일이어도 시기에 따라, 농장에 따라 당 함량이 달라지므로 제조하기 전 항상 당 확인을 해야 젤라띠에레가 원했던 단맛으로 정확하게 소르베또를 만들 수 있다.

① Scale

If the weight is not accurately observed when making gelato, it cannot be made according to the balanced recipe, which may give a different outcome than what the gelatiere wanted. I recommend using an electronic scale that goes up in increments of 1 gram and can weigh up to 30 kg.

② Hand Blender

It is a tool that completely blends the ingredients into one. A commercial-grade product with a high wattage rather than a household hand blender is recommended. It is a very important tool because the mixing result of the gelato mixture will vary depending on the performance of the hand blender, and it can affect the texture and taste of the completed gelato.

③ Thermometer

It is mainly used to measure the temperature when making the gelato base or mixture. It's also an essential tool for checking the temperature when making an infusing gelato.

④ Induction Cooktop/range

Gelato making does not require the use of fire, so there is no need to install a gas stove. However, it is good to have at least one induction cooktop to heat gelato that needs infusing, gelato that uses sub-ingredients (cooked rice, syrup, caramel cause, etc.), gelato base (white and yellow bases), gelato mixtures, etc.

⑤ Refractometer

It is an important tool used to check the sugar content of fruits used in sorbetto making. The sugar content of fresh fruits varies depending on the season and farms, even if they are the same fruits. Therefore, always check the sugar content before making the sorbetto with the exact sweetness the gelatiere wants.

Part 3.

Ingredients Used in Gelato

젤라또에 사용되는 재료들

THE MAPLE TREAT
100% pure
Maple Syrup
메이플 시럽
CANADA
GRADE A
AMBER, RICH TASTE
250 ml

1 Sugars

당류

모든 당류는 제각각의 단맛을 가지고 있다. 젤라또 세계에서 당류는 단맛을 포함해 젤라또의 질감과 서빙 온도에까지 영향을 미치는 주요 재료다. 따라서 레시피를 작성할 때에는 당류들의 감미도와 빙점강하력*을 조절해야 한다.

사용하는 당류에 따라 입 안에서 느껴지는 질감과 단맛 후미에서 올라오는 단맛 등이 다르다. 젤라띠에레의 경험과 선호하는 젤라또의 결과에 맞춰 당의 종류와 비율을 다르게 설정하여 더 부드럽고 달콤한 젤라또를 만들 수 있다.

많은 사람들이 당류는 단순히 단맛에만 영향을 미친다고 생각하지만 젤라또의 질감에 있어서도 중요한 영향을 미친다.

* 빙점강하력
물 100%는 0℃에서 얼지만 젤라또에는 수분과 고형분이 함께 있기 때문에 -2.5℃ 전후에서 얼기 시작한다. 이러한 현상을 '빙점강하력'이라 하며, 물질의 분자가 작을수록 분자의 숫자가 많아 더 높은 빙점강하가 일어난다. 가장 쉬운 예로 바닷물이 영하의 온도에도 얼지 않는 이유 역시 빙점강하 현상 때문인데, 바닷물 속의 소금 이온이 수분의 얼음 결정을 이루는 것을 방해하기 때문이다.
따라서 빙점강하력(PAC) 수치가 높은 당을 과도하게 사용할 경우 젤라또가 쇼케이스 안에서 얼지 않는다.

All sugars have their own sweet taste. When it comes to gelato, sugars play a significant role that affects not only the sweetness but also the texture and serving temperature of gelato. Therefore, you should adjust the 'sweetening power' and 'anti-freezing power*' when writing a recipe.

Depending on the sugars used, the sweetness you feel in the mouth, the sweetness that follows through the aftertaste, and the texture you feel in the mouth differ. A softer and sweeter gelato can be made by setting different types and ratios of sugars from the experience of the gelatiere and the result of their preferred gelato.

Many people think that sugars simply affect the sweetness, but they also play a vital part in the texture of gelato.

* Anti-freezing power
100% of water freezes at 0°C, but gelato starts to freeze around -2.5°C because it contains both water and solids. This phenomenon is called 'anti-freezing power,' the smaller the molecules of a substance, the greater the number of molecules, resulting in a higher anti-freezing point. As the easiest example, seawater does not freeze even at sub-zero temperatures because of the anti-freezing phenomenon due to salt ions in seawater preventing water from forming ice crystals.
Therefore, gelato will not freeze in the showcase if sugar with high anti-freezing power (PAC) value is used excessively.

종류 Types	**분자량** Molecular weight	**감미도(POD)** Sweetening power (POD*)	**빙점강하력(PAC)** Anti-freezing power (PAC*)	**고형분(%)** Solids (%)
설탕 Sugar	342	1.0	1.0	100
함수결정포도당 Dextrose	180	0.72	1.9	92
글루코스 시럽 DE 42-44 Glucose syrup DE 42-44	370	0.52	0.92	80
글루코스 시럽 파우더 DE 38-40 Glucose powder DE 38-40	750	0.23	0.45	96
꿀 Honey	190	1.3	1.9	80
전화당 Inverted sugar	190	1.25	1.9	70
말토덱스트린 DE 15-18 Maltodextrin DE 15-18	1,450	0.09	0.23	96
과당 Fructose	180	1.3	1.9	100
트레할로스 Trehalose	342	0.45	1.0	100
유당 Lactose	342	0.16	1.0	100
이눌린 Inulin	522	0.1	0.6	96
소금 Salt	58	-	5.9	100
알코올 Alcohol	46	-	7.4	-

DE : 포도당 당량 (Dextrose Equivalent)

*POD : 'Potere Dolcificante,' which means 'sweetening power' in Italian.
*PAC : 'Potere Anti-Congelante,' which means 'Anti-freezing power' in Italian.

전분의 가수분해 정도에 따라 전분당들의 감미도, 빙점강하력, 점성 등이 달라진다. 이러한 가수분해 정도를 가리키는 전 세계 공통의 표시가 바로 Dextrose Equivalent, 즉 'DE'이다.

DE값이 높을수록 전분보다 포도당의 비율이 높아 단맛이 강하고, DE값이 낮을수록 포도당보다 전분의 비율이 높아 점성이 강하다.

설탕

설탕은 우리에게 가장 익숙하고 인기 있는 대표적인 천연 감미료이다. 사탕수수와 사탕무를 착즙하고 결정화하여 원당을 만들고 원당을 정제하면 설탕이 된다. 정제 후 가열 과정을 거쳐 시럽화하여 재결정화하면 황설탕이 된다.

설탕은 필수 3대 영양소 중의 하나인 탄수화물의 원천으로, 신체 에너지의 연료로 작용되고 포도당 분자와 과당 분자가 결합된 이당류로 '자당'이라고도 불린다.

젤라또에 설탕을 첨가하여 풍미를 증진시키고, 감칠맛을 주기 위해 비정제당을 사용하기도 한다.

보통 너무 강한 단맛과 젤라또의 재결정화를 방지하기 위해 단독으로 설탕을 사용하지 않고 다른 당류와 혼합해서 사용한다.

함수결정포도당 (DE 100)

전분을 원료로 하는 대표적인 전분당인 포도당은 감미도가 설탕에 비해 낮고 차가운 온도에서도 용해가 빠른 당이다. 또한 설탕보다 빙점강하력이 강해 젤라또의 질감을 부드럽게 형성해줄 수 있지만 과도한 사용은 젤라또가 얼지 않고 구조를 약해지게 만든다. 또한 젤라또 표면이 광택이 나며 끈적해지는 현상이 일어나며 입 안에서 녹는 점을 낮게 만들어 더 차갑게 느껴지게 만든다. 그렇기 때문에 총 당량의 25% 이내에서 사용하는 것을 권장한다.

글루코스 시럽 (DE 30-64)

글루코스 시럽은 액상과 파우더, 두 가지 형태로 분류된다. 또한 전분의 가수분해 정도에 따라 DE값이 나뉜다. DE값이 높은 글루코스일수록 부드럽고 단맛이 강한 젤라또를 만들 수 있고, DE값이 낮은 글루코스 일수록 점성이 강하고 단맛이 강하지 않은 젤라또를 만들 수 있다. 일반적으로 젤라떼리아에서는 단맛의 강도와 점성의 강도가 비슷한 중간 DE값의 글루코스 시럽을 선호한다.

* 중간 DE값 : DE 38 ~ 44

꿀

천연감미료 중 가장 오래된 꿀은 감미뿐만이 아니라 건강에도 유익한 기능을 가지고 있다. 꽃으로부터 꿀을 모아 생산한 자연벌꿀의 경우 소량의 미네랄, 아미노산, 효소, 폴리페놀 등 항산화 효능과 면역 조절 효능을 가지고 있어 기능성 식품으로 분류되기도 한다. 꿀벌이 꽃가

Depending on the degree of hydrolysis of starch, the sweetness, anti-freezing power, and viscosity of starch sugars vary. A standard parameter used worldwide indicating the degree of hydrolysis is Dextrose Equivalent, or 'DE.'

The higher the DE value, the higher the ratio of glucose to starch, resulting in more robust sweetness. The lower the DE value, the higher the proportion of starch to glucose, resulting in stronger viscosity.

Sugar

Sugar is one of the most familiar and typical natural sweeteners. Sugar canes and beets are juiced and crystallized to make raw sugar, which is refined into sugar. After refining, it is made into syrup through a heating process and recrystallized to brown sugar.

Sugar is a source of carbohydrates, one of the three essential nutrients, and serves as a fuel for body energy. It's also called 'sucrose,' a disaccharide composed of glucose and fructose molecules.

Unrefined sugar is sometimes used to enhance flavor and give a tasty flavor by adding sugar to gelato.

Usually, the sugar is not used alone to prevent too intense sweetness and recrystallization of gelato but mixed with other sugars.

Dextrose (DE100)

Dextrose, a typical starch sugar made from starch, has a lower sweetness than sugar and dissolves quickly, even at cold temperatures. Also, it has a stronger anti-freezing point than sugar, so that it can make a soft textured gelato. Excessive use will cause the gelato not to freeze and weaken its structure. Also, gelato's surface becomes glossy and sticky, and it lowers the melting point in the mouth, making it feel colder. Accordingly, it is recommended to use within 25% of total sugar weight.

Glucose Syrup (DE 30-64)

Glucose syrup comes in two forms: liquid and powder type. Also, it's divided into DE values according to the degree of hydrolysis of starch. The higher the DE value, the softer and sweeter the gelato; the lower the DE value, the harder and less sweet the gelato will be. In general, a gelateria prefers glucose syrup with a medium DE value that has a similar sweetness and viscosity.

* Regular conversion DE value : DE 38~44

Honey

Honey is the oldest form of natural sweetener that has not only sweetness but also a few health benefits. Natural honey collected from flowers is classified as a functional food because it has antioxidant and immune-modulating effects, such as small amounts of minerals, amino acids, enzymes, and polyphenols. The honey

루가 아닌 설탕물을 인위적으로 먹고 얻어낸 꿀은 자연꿀이 아닌 사양벌꿀이라 부른다.

자연꿀은 밀원에 따라 분류가 되고 일반적으로 아카시아 꿀, 밤 꿀, 잡화(야생화) 꿀 등이 있다.

꿀은 밀원에 따라 느껴지는 맛과 향이 다르기에 젤라또에 잘 사용하면 매력적인 맛을 표현할 수 있다.

collected from bees that consumed sugar water than pollen is not natural honey but is called imitation honey.

Natural honey is classified according to its nectar source and can generally include acacia honey, chestnut honey, and mixed (wild) flower honey.

Honey has a different taste and aroma depending on the nectar source, so you can create an attractive taste when used well.

전화당 (트리몰린)

전화당은 설탕을 산이나 효소로 가수분해하여 만든 당으로 설탕보다 단맛이 강하고 빙점감하력이 높고 수분의 흡습성이 좋다. 꿀과 동일한 역할을 하지만 꿀 특유의 향과 알레르기 유발 성분이 없어 꿀 대체제로 사용되기도 한다.

Inverted Sugar

Inverted sugar is made by hydrolysis of sugar with acids or enzymes. It is sweeter than sugar, has high anti-freezing power, and has good moisture absorption. It has the same role as honey, but it's also used as a honey substitute because it does not have the unique aroma and allergy-causing components of honey.

말토덱스트린 (DE 9-19)

전분당 중 DE값이 20 이하인 당들은 말토덱스트린이라 부른다. 낮은 감미도와 낮은 빙점강하력으로 인해 젤라또에서 당류로의 역할보다는 점성을 강하게 부여하는 역할을 하며, 안정제를 서포트하는 용도로 주로 사용된다. 찰기 있는 식감이 강한 젤라또를 선호하는 젤라떼리아에서는 말토덱스트린을 사용함으로 젤라또에 찰기를 보다 더 강하게 부여해준다.

Maltodextrin (DE 9-19)

Among the starch sugars, sugars with a DE value of less than 20 are called maltodextrin. Due to its low sweetening power and low anti-freezing power, it plays a role in imparting a strong viscosity rather than a role as a sugar in gelato and is mainly used for supporting stabilizers. A gelateria who prefers a gelato with a glutinous texture uses maltodextrin to give a firmer glutinous texture.

과당

과일과 전분에 존재하는 과당은 다른 당류에 비해 단맛이 강하고 빙점강하력 역시 높다. 혈당 지수가 낮고 인슐린 분비를 직접적으로 증가시키지 않아 당뇨 환자에게 적합하다. 하지만 과다한 복용은 간을 손상시키고 비만을 초래할 수 있다. 또한, 너무 강한 단맛으로 젤라또에서 다른 재료의 맛이 충분히 느껴지지 않을 수 있다.

Fructose

Fructose exists in fruits and starches and has a more substantial sweet taste and higher anti-freezing powder than other sugars. It has a low glycemic index and does not directly increase insulin secretion, making it suitable for diabetic patients. However, too much consumption can damage the liver and lead to obesity. Also, the taste of other ingredients in gelato may not be able to fully perceived due to its intense sweetness.

트레할로스

일본에서 개발된 당으로 버섯, 효모로부터 만들어진 당류이다. 설탕과 비교했을 때 절반 이상 낮은 감미도를 가지고 있지만 설탕과 동일한 빙점강하력을 가지고 있어 단맛을 선호하지 않는 곳에서 설탕과 혼합하여 사용하며, 비건 젤라또를 만들 때에도 사용된다.

Trehalose

It is a sugar developed in Japan and made from mushrooms and yeast. Compared to sugar, it has less than half the sweetening power but has the same anti-freezing power as sugar. So it is mixed with sugar in recipes where sweetness is not preferred and is mainly used when making a vegan gelato.

유당

유제품에 함유된 당류로 용해성이 낮아 많이 사용되면 젤라또에서 거친 입자감(모래화 현상)이 느껴진다. 우유 안에는 5.5% 정도의 유당이 함유되어 있다.

Lactose

It's a sugar contained in dairy products and has low solubility. So, if you use it a lot, you can feel the rough particles (sand-like phenomenon) in gelato. Milk contains about 5.5% of lactose.

이눌린

이눌린은 치커리 뿌리, 돼지감자로부터 얻어지는 다당류로 식물성 식이섬유이다. 젤라또의 단맛과 빙점강하력에는 영향을 미치지 않아 고형분이 낮은 젤라또나 소르베또에 사용하면 고형분 함량을 높여 질감을 향상시키는 데 도움을 준다.

Inulin

Inulin is a polysaccharide obtained from chicory root and Jerusalem artichoke and is a vegetable dietary fiber. It does not affect the sweetness and anti-freezing power in gelato, and when used in gelato or sorbetto with low solids, it helps to improve the texture by increasing the solids content.

대체당

당뇨, 고칼로리 섭취 등의 이슈로 최근 설탕을 대체하는 당으로 만든 식품이 많이 나오고 있다. 무가당, 저칼로리 젤라또를 만들 때 대표적으로 사용되는 대체당은 천연당, 당알코올, 천연 감미료, 합성 감미료로 분류된다. 혈당지수(GI)와 칼로리가 낮지만 과도한 사용은 복통을 유발하고 당 알코올의 경우 특유의 화한 청량감을 느낄 수 있다.

몸무게 1kg당 당 알코올류 1g 이상의 섭취는 권장하지 않는다.

예) 몸무게 70kg의 사람이 섭취할 수 있는 당 알코올류의 양은 70g 이하이다.

* 혈당지수(GI : Glycemic Index)

'혈당지수'란 식품을 섭취했을 때 혈당의 상승률을 수치로 나타낸 것이다. 혈당지수가 높은 식품은 탄수화물이 빠르게 분해되어 혈당이 오르고, 혈당지수가 낮은 식품은 천천히 분해되어 혈당이 느리게 올라간다.

◆ 무가당 젤라또는 설탕이 들어간 젤라또와 비교했을 때 칼로리가 낮고 혈당을 과하게 높이지 않는다는 의미이다. 무가당 젤라또라고 해서, 당이 들어가지 않았다고 생각하는 것은 잘못된 접근 방식이다. 열량이 높지 않고 단맛이 강하지 않아 되려 과도한 섭취를 하게 되면 오히려 건강에 해가 될 수 있다.

Sugar Substitutes

Due to issues such as diabetes and high-calorie intake, food products made with sugar substitutes have recently emerged. Sugar substitutes typically used when making sugar-free and low-calorie gelatos are classified into natural sugars, sugar alcohols, natural sweeteners, and synthetic sweeteners. Although the glycemic index (GI) and calories are low, excessive use can cause abdominal pain. As for sugar alcohol, it can give you a unique, mint-like refreshment.

It is not recommended to consume more than 1 gram of sugar alcohol per 1 kilogram of body weight.

E.g. The amount of sugar alcohol a 70 kg person can consume is less than 70 g.

* Glycemic Index (GI)

'The glycemic index' is a numerical value indicating the rate at which blood sugar rises when food is consumed. Foods with a high glycemic index break down carbohydrates quickly and raise blood sugar, while foods with a low glycemic index break down slowly and raise blood sugar slowly.

◆ Sugar-free gelato means that it has fewer calories than when sugar is used and doesn't excessively raise blood sugar levels. It is a wrong approach to assume that sugar-free gelato does not contain sugar. Excessive intake can be harmful to health because it is not high in calories and doesn't taste too sweet.

대체당의 종류 Types of sugar substitutes	**분자량** Molecular weight	**감미도(POD)** Sweetening power (POD*)	**빙점강하력(PAC)** Anti-freezing power (PAC*)	**고형분(%)** Solids (%)
에리스리톨(E968) Erythritol (E968)	122	0.7	2.8	95
이소말트(E953) Isomalt (E953)	342	0.5	1.0	95
말티톨(E965) Maltitol (E965)	344	0.9	1.0	95
만니톨(E421) Mannitol (E421)	182	0.7	1.9	95
솔비톨(E420) Sorbitol (E420)	182	0.6	1.9	95
자일리톨(E967) Xylitol (E967)	152	1.0	2.25	95
스테비아 Stevia	967	300	-	95
아가베 시럽 Agave syrup	190	1.4	1.8	68
메이플 시럽 Maple syrup	380	0.6	0.9	67

에리스리톨 (E968)

과일, 버섯 등에서 발견되는 에리스리톨은 체내에 약 10%만 흡수되고 나머지는 체외로 배출되는 감미료이다. 무가당 젤라또를 만들 때

Erythritol (E968)

Erythritol is found in fruits and mushrooms, and it is a sweetener that only absorbs about 10% of it in the body and excretes the rest. It can be used instead of glucose when making sugar-free

포도당 대신 사용할 수 있는 감미료이지만 빙점강하력이 포도당보다 높기 때문에 주의해야 한다.

gelato, but be aware because it has higher anti-freezing power than glucose.

이소말트 (E953)

사탕무에서부터 발견되는 이소말트는 설탕과 비슷한 특성을 가지고 있는 감미료이다. 설탕 공예를 할 때 주로 사용되며, 열 안전성이 높아 고온에서 가열해도 캐러멜화되지 않고 0°C 이하의 온도에서도 제품이 잘 유지되는 특성을 가지고 있다.

Isomalt (E953)

Isomalt is a sweetener found in beets with properties similar to sugar. It's usually used in sugar crafts and does not caramelize even at high temperatures due to its high thermal stability. Also, it has a characteristic that keeps the product stable even at a temperature below 0℃.

말티톨 (E965)

과일과 채소에서 발견할 수 있는 말티톨은 설탕과 거의 같은 단맛을 가지고 있어 무가당 젤라또를 만들 때 설탕 대신 사용할 수 있다. 하지만 과한 섭취는 복통을 유발한다.

Maltitol (E965)

Maltitol can be found in fruits and vegetables. It has almost the same sweetness as sugar, so it can replace sugar when making sugar-free gelato. However, excessive consumption can cause abdominal pain.

만니톨 (E421)

과일과 해조류에서 추출되는 만니톨은 (예를 들어) 곶감이나 건조된 다시마 겉면의 흰색 가루이다. 단맛을 내면서도 설탕과는 다르게 습기를 먹지 않아 사탕 표면을 코팅하는 용도로도 사용된다.

Mannitol (E421)

Mannitol is extracted from fruits and seaweeds, and it's the white powder on the surface of dried persimmons or dried kelp, for example. It has a sweet taste, but unlike sugar, it does not absorb moisture, so it is also used to coat the surface of candies.

솔비톨 (E420)

과일에서 발견되는 솔비톨은 흡습성이 좋고 향을 잘 유지해주는 역할을 한다. 대표적으로 프룬에 풍부하게 함유되어 있고, 설탕보다 감미도가 낮고 같은 중량 기준 칼로리가 40% 더 낮다.

Sorbitol (E420)

Sorbitol is found in fruits, and it's hygroscopic and helps to retain flavor. Typically, it is abundantly contained in prunes, is less sweet than sugar, and has 40% less calories by weight.

자일리톨 (E967)

핀란드 자작나무에서 발견되는 자일리톨은 치아 건강에 유의한 역할을 하여 주로 껌의 감미료로 사용된다.

Xylitol (E967)

Xylitol is found in Finnish birch trees and plays a significant role in dental health, and is primarily used as a sweetener for chewing gums.

스테비아

중남미 지역의 허브로 스테비아 잎이나 줄기 속 단맛을 내는 물질을 추출해서 만든 감미료이다. 탄수화물을 섭취하면 체내에서 포도당으로 분해가 되어 혈당 수치를 높이는데, 스테비아는 포도당으로 분해되지 않아 체내에 흡수되지 않고 혈당에 영향을 미치지 않는다. 설탕의 300배 이상의 단맛을 가진 감미료로 스테비아 0.01g = 설탕 3g과 같은 단맛을 낸다.

Stevia

Stevia is an herb native to Central and South America and is a sweetener made by extracting a sweet-tasting substance from the leaves or stems. Carbohydrates are broken down into glucose to raise blood sugar levels. But stevia does not break down into glucose, so it is not absorbed into the body and does not affect blood sugar levels. It is a sweetener 300 times sweeter than sugar, and 0.01 grams of stevia equals 3 grams of sugar.

아가베 시럽

멕시코에서 자라는 용설란에서 추출하여 만든 시럽으로 철분, 칼슘, 미네랄 등 영양소가 풍부하다. 과당으로 구성되어 혈당지수(GI)가 낮지만, 설탕보다 25~30% 더 달다.

Agave syrup

It is a syrup made from extracts of agave grown in Mexico and is rich in nutrients such as iron, calcium, and minerals. It consists of fructose and has a low glycemic index (GI) but is 25~30% sweeter than sugar.

메이플 시럽

단풍나무 수액을 끓인 뒤 증기를 모아 식혀 만든 시럽으로 비타민과 미네랄이 풍부하다.

Maple syrup

It is a syrup made by boiling the sap of a maple tree and collecting steam to cool it. It's rich in vitamins and minerals.

2 Dairy Products

유제품

젤라또에서 느껴지는 고유의 풍미와 크리미함은 지방 고형분에서부터 온다. 젤라또에 사용되는 여러 재료가 지방을 함유하고 있는데, 그 중 가장 대표적이면서도 높은 비율을 차지하는 지방은 유제품으로부터 오는 유지방이다.

The unique flavor and creaminess of gelato come from solid fat. There are several ingredients used in gelato that contains fat. Among them, milk fat from dairy products is the most representative and takes a high percentage.

우유

젤라또에 사용되는 핵심 재료 중 하나인 우유는 60% 이상의 높은 첨가 비율을 차지한다. 따라서 어떤 우유를 사용하느냐에 따라 최종적으로 완성되는 젤라또의 맛에 영향을 미칠 수 있다.

유지방 함량에 따라 일반 우유(3~4%), 저지방 우유(1.5~2%), 무지방 우유(0.1%)로 나뉜다.

부피와 중량이 동일한 물과 다르게 우유는 87.5%만 수분이고 나머지는 고형분이기 때문에 액상이어도 부피와 중량이 똑같을 수 없다. (물 1,000ml = 1,000g이지만, 우유 1,000ml = 1,030g이다.)

우유 1L 기준으로 젤라또 레시피를 작성하면 많은 양의 우유를 계량하지 않고 바로 사용할 수 있다.

예) 레시피상 우유가 5L 들어가면 1L짜리 우유를 따로 계량하지 않고 우유 5개를 바로 사용하면 된다.

Milk

Milk is one of the key ingredients used in gelato and has a high milk fat ratio of over 60%. Therefore, the taste of the completed gelato can be affected depending on the milk used.

Depending on the milk fat content, it is divided into regular milk (3~4%), low-fat milk (1.5~2%), and non-fat milk (0.1%).

Unlike water, which has the same volume and weight, milk has only 87.5% of water, and the rest is solids, so even if it's in a liquid form, the volume and weight cannot be the same. (1,000 ml of water weighs 1,000 g, but 1,000 ml of milk is 1,030 g.)

If you create a gelato recipe based on 1L of milk, you can use it immediately without measuring a large amount of milk.

E.g. If the recipe contains 5L of milk, you can instantly open and use five packs without having to measure them individually. (If your milk is 1L per pack.)

원유

우유 고유의 진한 풍미를 중요시 여기는 젤라띠에레는 목장에서 착유한 살균 전 원유로 만드는 젤라또를 선호한다. 하지만 국내에서는 일반 소비자가 원유를 구입할 수 없으므로, 목장에서 원유를 사용해 만든 밀크팜 베이스나 목장에서 살균 후 판매하는 단일 목장 우유를 선택하여 젤라또를 만들 수 있다.

Raw Milk

Gelatiere, who values the unique rich flavor of milk, prefers the gelato made from raw milk that has been milked on a ranch before pasteurization. However, domestic consumers cannot purchase raw milk. Hence, you can select a milk base made from raw milk at the ranch or single ranch milk sold after pasteurization at the ranch.

시유

한 곳의 목장에서 착유하여 유통되는 단일 목장 우유와 다르게 시유는 여러 목장의 원유를 우유 공장에 집유해 와서 하나의 표준화된 제품으로 판매된다. 소위 우리가 쉽게 구입하여 사용할 수 있는 브랜드 우유이다. 브랜드와 제품마다 원유를 살균하는 온도가 다르기에 마셨을 때 느껴지는 풍미가 다르다.

Commercial Milk

Unlike milk from a single farm that is milked and distributed at one farm, commercial milk is sold as a standardized product after milk from multiple farms is collected at a milk factory. It is so-called branded milk that we can easily buy and use. Each brand and product has a different temperature for sterilizing raw milk, so the flavor you taste when you drink may differ.

멸균우유

멸균 과정을 거쳐 우유 팩을 열기 전 실온에서 장기 보관이 가능하여 냉장 공간이 협소한 젤라떼리아에서 사용 가능한 우유이다.

UHT Milk

This milk can be stored for a long time at room temperature before opening due to a pasteurization process and can be used in gelaterias, where refrigeration space is limited.

환원유

물과 탈지분유를 혼합하여 우유와 유사한 조성으로 만든 환원유는 식품 유형이 우유가 아닌 가공유로 분류된다. 우유 본연의 진하고 고소한 맛은 없지만 젤라또에서 우유의 뒷맛 없이 깔끔한 맛이 나는 것을 선호하는 젤라띠에레에게는 적합한 제품이다.

Reconstituted Milk

Reconstituted milk is made by mixing water and skim milk powder to have a composition similar to milk, and it's classified as processed milk rather than milk as a food type. It does not have the original rich and savory taste of milk, but it's a suitable product for a gelatiere who prefers a clean taste without the aftertaste of milk in gelato.

생크림

젤라또에서 지방 고형분을 채우기 위한 핵심 재료는 유지방 함량 35~38%의 생크림이다. 일반적으로 국내산 동물성 생크림 또는 수입산 동물성 휘핑크림을 사용하여 젤라또의 유지방 풍미를 높여준다. 국내산 동물성 생크림은 깔끔한 유지방 풍미를 가지고 있고, 수입산 동물성 휘핑크림은 진하고 고소한 유지방 풍미를 가지고 있다.

분유

우유의 수분을 증발시키고 농축하여 만든 분말을 분유라고 하며, 유지방 함량이 26%인 것을 전지분유, 우유에서 유지방을 완전히 분리하여 만든 것을 탈지분유라 한다.

분유는 젤라또의 유단백질 함량을 높여 뼈대를 형성하는 역할을 한다. 우유와 생크림으로 젤라또에 필요한 유지방을 채울 수 있지만, 유단백질은 채울 수 없기 때문에 유지방이 없는 탈지분유를 사용하여 충족시켜준다.

저가형 아이스크림은 물과 탈지분유를 혼합해서 우유 맛을 만들기 때문에, 젤라또에도 탈지분유가 들어간다고 하면 우유 맛을 인위적으로 만들기 위해 사용한다고 생각하는 소비자들이 있다. 하지만 젤라또에 탈지분유의 역할은 우유 맛을 내기 위해서가 아니라 유단백질을 높여 공기를 잡아 젤라또의 구조를 개선하고 오버런을 증가시켜 입안에서 젤라또가 따뜻하게 느껴지게 하는 것이다. 또한, 젤라또의 녹는 속도를 늦추는 역할로 사용되기도 한다.

✦ 우유를 사용할 수 없는 환경에서는 분유와 물을 혼합하여 환원유를 직접 만들어 젤라또 제조를 하는 경우도 있다. 예를 들어 물 900g + 전지분유 130g = 1,030g으로 일반우유를 대체할 수 있다.

버터

우유의 지방을 분리하여 응고시켜 만든 유제품인 버터는 젤라또에 생크림 대신 사용되기도 한다. 생크림보다 유지방 함량이 높아 적은 양으로도 젤라또에 필요한 지방을 충족시킬 수 있지만, 버터 특유의 풍미로 인해 모든 젤라또에 사용하기에는 맛이 어울리지 않을 수 있다.

요거트

우유에 유산균을 배양하여 만들어지는 발효유이다. 젤라또 제조기에 요거트를 만드는 기능이 있다면 직접 만든 요거트로 요거트 젤라또를 만들 수 있다.

✦ Carpigiani사 기준으로 Ready Chef, Maestro HE 가능

만드는 방법은 비터에서 스크래퍼를 뺀 제조기에 우유를 넣고 요거트 모드를 사용한다. 우유가 90~95°C까지 가열되었다가 42°C로 냉각되면 무가당 플레인 요거트 또는 유산균을 첨가한 후 6~10시간 발효시킨다. 완성된 요거트는 냉장에서 2주간 보관하며 사용이 가능하다.

✦ 우유 1L 기준 무가당 플레인 요거트 100g. 유산균을 사용할 경우 각 회사의 권장량을 따른다.

Cream

The key ingredient to fill the fat solids in gelato is cream with a milk fat content of 35~35%. Generally, domestic or imported dairy cream is used to enhance the milk fat flavor of gelato. Domestic dairy cream has a fresh milk fat flavor, and imported cream has a rich and savory milk fat flavor.

Milk Powder

Powder made by evaporating and concentrating the moisture of milk is called powdered milk. Milk powder with a milk fat content of 26% is called whole milk powder, and milk powder made by completely separating milk fat from milk is skim milk powder.

Milk powder increases the milk protein content of gelato and plays a role in forming the structure. Milk and cream can suffice the milk fat needed for gelato. But, there is not enough milk protein, so skim milk powder without milk fat is used to compensate for it.

Since low-priced ice creams use water and skim milk powder for the milk taste, some customers think that skim milk powder is used in gelato to add the milk taste artificially. However, the role of skim milk powder is not to add taste but to improve the structure of gelato by capturing air by increasing milk protein and increasing the overrun so the gelato feels warm in the mouth. It is also used to slow down the melting of gelato.

✦ In environments where milk cannot be used, there are cases in which gelato is made by mixing milk powder and water to make reconstructed milk. For example, you can replace 1,030 grams of regular milk by mixing 900 grams of water and 130 grams of whole milk powder.

Butter

Butter is a dairy product made by separating and coagulating the fat of milk, which is sometimes used instead of cream in gelato. It has a higher milk fat content than cream, so even a small amount can satisfy the fat needed for gelato. However, due to its unique flavor, it may only be suitable for some gelato flavors.

Yogurt

Yogurt is fermented milk made by culturing lactic acid bacteria in milk. If your gelato maker has a yogurt-making function, you can make yogurt gelato with your own yogurt.

✦ As for Carpigiani, 'Ready Chef' and 'Maestro HE' are capable.

To make yogurt, put milk in the maker with the scraper removed from the paddle and use the yogurt mode. When the milk heats to 90~95°C and cools to 42°C, unsweetened plain yogurt or lactic acid bacteria are added and fermented for 6~10 hours. The completed yogurt and be stored and used in the refrigerator for two weeks.

✦ Use 100 grams of unsweetened plain yogurt per 1 liter of milk. When using lactic acid bacteria, follow the recommended dosage of the company.

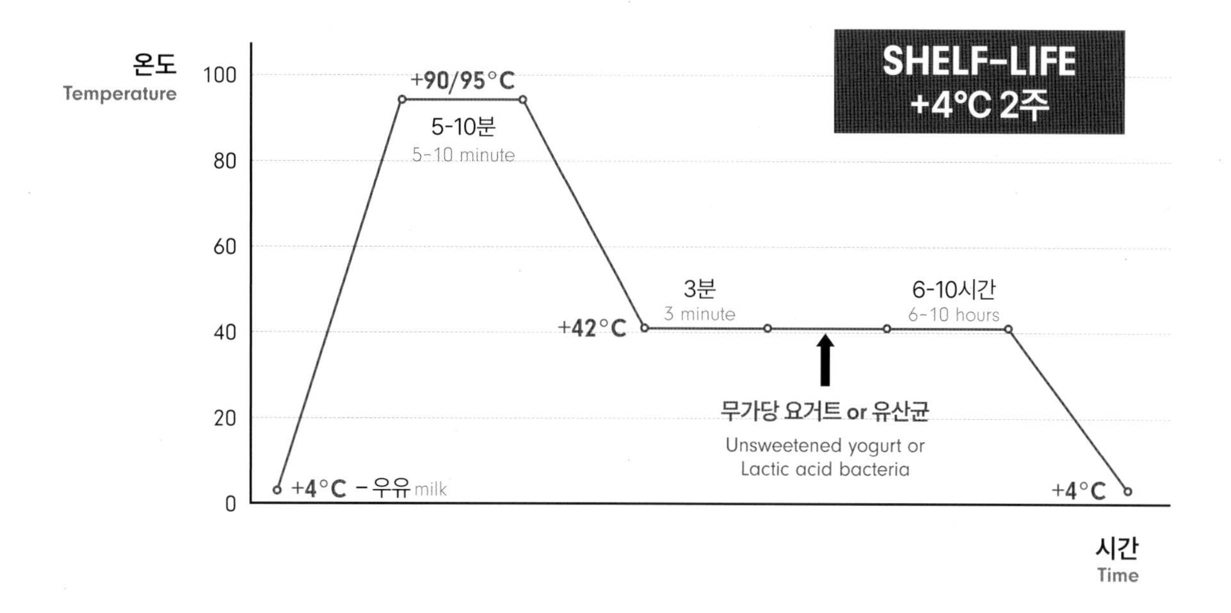

치즈

우유에 함유된 유단백질 분해 효소제인 렌닛이나 산을 넣어 응고시켜 커드를 형성한 것이 치즈이다.

커드의 크기에 따라 수분 함량이 달라지는데 크게 생치즈, 연성치즈, 경성치즈로 분류된다. 발효와 숙성을 거치지 않고 만들며, 수분 함량이 가장 높은 생치즈는 모짜렐라, 부라타 등이 있다. 수분 함량이 55% 이상인 연성치즈는 브리, 까망베르 등이 있고, 가장 수분이 적고 지방과 단백질 함량이 높은 경성치즈는 파르미지아노 레지아노, 그라나 파다노 등이 있다.

푸른 곰팡이에 의해 숙성되는 반경성치즈는 고르곤졸라, 로크포르 등이 있다.

우유에서 분리한 크림을 사용해 발효 과정 없이 만들어지는 대표적인 치즈는 크림치즈와 마스카르포네가 있다.

리코타는 치즈를 만들고 남은 유청에 산을 첨가하여 다시 재가열해 유청 단백질이 분리되어 뭉쳐 만들어지는 유청치즈이다. 이탈리아에서는 리코타가 커드가 아닌 부산물로 만들어지기 때문에 치즈로 정의하지 않고 유제품으로 정의한다.

디저트 젤라또에는 치즈 고유의 향이 강하지 않은 크림치즈, 마스카르포네, 리코타가 주로 사용되고, 세이보리 젤라또에는 그 외의 치즈들이 사용된다.

Cheese

Cheese is made by coagulating rennet or acid, which is a milk proteolytic enzyme contained in milk, to form a curd.

Moisture content varies depending on the size of the curd, and it is largely classified into fresh, soft, and hard cheese. Mozzarella and burrata are fresh cheeses that are made without fermentation and aging and have the highest moisture content. Soft cheeses with a moisture content of 55% or more include Brie and Camembert, and hard cheeses with the lowest moisture, high fat, and protein content include Parmigiano Reggiano and Grana Padano.

Semi-hard cheeses aged by blue mold include Gorgonzola and Roquefort.

Representative cheeses made without fermentation using cream separated from milk include cream cheese and mascarpone.

Ricotta is a whey cheese made by adding acid to the whey left over from making cheese and reheating to separate and coagulate the whey proteins. In Italy, ricotta is not defined as cheese but as a dairy product, as it is made from a by-product rather than curd.

Cream cheese, mascarpone, and ricotta, which do not have a strong cheese flavor, are mainly used for dessert gelato, and other cheeses are used for savory gelato.

3 Stabilizers
안정제

젤라또에서 가장 소량으로 사용되는 재료이지만, 젤라또의 질감을 결정하는 중요한 역할을 하는 식품첨가물이다. 원하는 완성도에 따라 사용되는 안정제와 용량이 달라진다.

안정제는 수분을 흡수하고 얼음 결정을 작게 만들며 점성을 부여하는 증점제와, 지방과 수분의 안정적인 결합을 도와주는 유화제가 있다.

Although it's used in the smallest amount in gelato, it is a food additive that plays a vital role in determining the texture of gelato. The amount used and its dosage vary depending on the degree of completion you want.

Stabilizers include a thickener that absorbs moisture, makes ice crystals smaller, and provides viscosity, and an emulsifier that helps stably combine fat and water.

Thickener 증점제

증점제의 경우 젤라띠에레가 원하는 검류를 직접 구입하여 검류의 특성과 사용량을 이해한 후 배합해서 사용해야 한다. 하나의 증점제로 완벽한 젤라또의 질감을 만들 수 없기에 서로 다른 특성을 가진 증점제를 혼합하여 서로의 결점을 보완해주어야 한다.

When using the thickener, gelatiere should purchase the desired gum and understand the characteristics and dosage of the gum before mixing and using it. Because one type of thickener cannot make the perfect texture, different thickeners with different characteristics must be combined to compensate for each other's shortcomings.

아가아가 (E406)

아시아 해안의 붉은 해조류에서 얻어지는 다당류로 겔화 능력이 뛰어나다. 온도 변화가 잦아도 안정화 기능이 뛰어나고 온수에서 작용한다. 젤라떼리아에서 많이 사용되는 증점제는 아니지만 알긴과 혼합해서 사용하는 것이 이상적이다. 한국에서는 한천이라고도 불리며 0.1~0.5% 비율로 사용할 수 있다.

Agar-agar (E406)

It is a polysaccharide obtained from red seaweed from the Asian coast and has excellent gelling ability. It has an excellent stabilizing function and works with hot water even with frequent temperature changes. It's not a commonly used thickening agent in a gelateria, but it's ideal to use it in combination with algin. It's called 'hancheon (agar)' in Korean and can be used between 0.1~0.5%.

알긴 (E400-405)

영국, 노르웨이 등 북해도 해조류에서 추출되는 알긴산은 수분을 뛰어나게 흡수하는 능력을 가지고 있다. 알긴은 알긴산(E400), 알긴산소듐(E401), 알긴산 칼륨(E403), 알긴산 칼슘(E404), 알긴산 프로필레글리콜(E405)로 분류된다. 온수에 작용되고 알긴산소듐의 경우 냉수에도 사용 가능하다. 젤라떼리아에서는 젤라또의 점성과 오버런에 도움을 주는 알긴산소듐(E401)을 선호하며 0.15~0.25% 비율로 사용할 수 있다.

Algin (E400~405)

Alginic acid is extracted from seaweed in the Northern Seas of England and Norway and has an excellent ability to absorb moisture. It is classified into alginic acid (E400), sodium alginate (E401), potassium alginate (E403), calcium alginate (E404), and propylene glycol alginate (E405). It activates in hot water, and sodium alginate can be used in cold water. Sodium alginate (E401), which helps with viscosity and overrun, is preferred in the gelateria and can be used between 0.15~0.25%.

CMC (E466)

식물의 셀룰로오스에서 추출한 CMC는 무미, 무취로 냉수와 온수 겸용으로 사용되며 부분적으로 유화의 특성도 있다. 다양한 온도에서

CMC (E466)

CMC extracted from plant cellulose is tasteless and odorless. It can be used in both cold and hot water and has some emulsifying capability. Although it helps gelato to be stable

젤라또가 안정성을 가지지만 단독으로 사용하지 않고 카라기난과 함께 0.1~0.25%로 혼합해 사용하는 것이 좋다. 초콜릿 젤라또에 사용하면 카카오 입자의 분리를 막아 보다 더 효과적으로 안정화시킬 수 있다.

at various temperatures, it is recommended to mix it with carrageenan at 0.1~0.25% instead of using it alone. It can prevent the separation of cacao particles and stabilize them more effectively when used in chocolate gelato.

카라기난 (E407)

아일랜드와 영국 해안에서 자라는 붉은 해조류에서 추출한 카라기난은 온수에서 작용한다. 젤라또의 질감과 녹는 속도를 늦추기 위해 다른 증점제와 함께 사용된다.(수입산 휘핑크림에 카라기난이 첨가되는 것을 볼 수 있다.) 초콜릿 젤라또에도 잘 어울리며 0.2~0.5% 비율로 사용할 수 있다.

Carrageenan (E407)

Carrageenan is extracted from red algae that grows on the coasts of Ireland and England and works with hot water. It is used with other thickeners to slow down the melting and for the texture of gelato. (You can find that carrageenan is added to imported creams.) It works well with chocolate gelato and can be used between 0.2~0.5%.

로커스트빈검 (E410)

시칠리아와 지중해 지역의 캐롭 나무 씨앗에서 얻어지는 로커스트빈검은 무게에 비해 100배 이상의 높은 수분 흡수력을 가져 부드러운 질감과 녹는 속도를 늦춰 젤라또에 가장 많이 사용되는 증점제이다. 온수에서 작용되고 증점의 최대 효과를 얻기 위해서는 숙성 시간이 필요하다. 구아검이나 카라기난과 함께 사용될 때 더 좋은 결과를 만들어내고 0.3~0.5% 비율로 사용할 수 있다.

Locust Bean Gum (E410)

Locust bean gum is collected from the seeds of the carob tree in Sicily and the Mediterranean region. It is the most commonly used thickener for gelato because it has a high moisture absorption capacity of more than 100 times its own weight, has a soft texture, and slows down the melting speed. It activates in hot water and requires aging time to achieve the maximum thickening effect. It produces better results when used with guar gum or carrageenan and can be used between 0.3~0.5%.

구아검 (E412)

인도와 파키스탄의 구아 식물의 씨앗에서 추출되는 구아검은 냉수에서 작용되어 소르베또에 주로 사용된다. 냉수에서 작용되는 증점제는 온수에서도 활동하므로 구아검은 로커스트빈검과 함께 혼합되어 젤라또에도 가장 많이 사용되는 증점제이다. 0.3~0.4% 비율로 사용할 수 있다.

Guar Gum (E412)

Guar gum is derived from the seeds of the guar plant in India and Pakistan. It activates in cold water and is used primarily in sorbetto. Since the thickener that works in cold water can also be used in hot water, guar gum mixed with locust bean gum is the most commonly used thickener for gelato. It can be used between 0.3~0.4%.

타라검 (E417)

페루나 볼리비아 등 남미 국가의 카이살피니아 스피노사 식물의 씨앗에서 추출되는 타라검은 냉수에서 작용되며 동결과 해동의 안정성이 뛰어나 젤라또에 얼음 결정이 형성되는 것을 방지한다. 0.3~0.5% 비율로 사용할 수 있다.

Tara Gum (E417)

Tara gum is extracted from the seeds of the Caesalpinia Spinosa plant in South American countries such as Peru and Bolivia. It works in cold water, has excellent freezing and thawing stability, and prevents ice crystals from forming in gelato. It can be used between 0.3~0.5%.

잔탄검 (E415)

식물에서 얻어진 박테리아에 탄수화물을 주입 후 발효시켜 만든 잔탄검은 냉수에서 작용되며 수분 흡습성이 아주 뛰어나고 젤라또가 입안에서 차갑게 느껴지지 않는 역할을 한다. 로커스트빈검과 함께 사용할 수 있고 0.05~0.08% 비율로 소량 사용해야 한다.

Xanthan Gum (E415)

Xanthan gum is made by injecting carbohydrates into bacteria obtained from plants and fermenting them. It works in cold water, has excellent moisture absorption, and helps gelato not feel cold in the mouth. It can be used with locust bean gum and should be used in small amounts between 0.05~0.08%.

펙틴 (E440)

과일과 감귤류 껍질에서 추출되는 펙틴은 산성에 아주 강한 작용을 하고 온수에서 작용한다. 펙틴 HM(High-methoxy)과 펙틴 LM(Low-methoxy)으로 나뉘며 HM은 주로 고당도 잼이나 젤리를

Pectin (E440)

Pectin is extracted from fruits and citrus peels, works well with acid, and activates in hot water. There are two types of pectins; HM (high methoxyl) and LM (low methoxyl) pectin. HM pectin is mainly used when making high-sugar jams or jellies, and LM

만들 때 사용되고, LM은 젤라떼리아에서 0.5~1% 비율로 사용할 수 있지만 다른 증점제와 혼합해 사용하는 것을 권장한다.

pectin can be used in gelateria between 0.5~1%, but mixing with other thickeners is recommended.

살렙

터키와 이란의 난초 구근을 말려 만들어지는 살렙은 터키 아이스크림인 돈두르마의 쫀득하게 씹히는 식감에 영향을 미치는 증점제이다.

Salep

Salep is made from dried orchid bulbs from Türkiye and Iran. It is a thickener that gives the chewy texture of Dondurma, a Turkish ice cream.

Emulsifier 유화제

젤라또에 존재하는 물과 지방을 안정적이게 결합하고 지방과 단백질의 상호작용을 촉진한다. 또한 온도 변화에 저항할 수 있는 역할과 오버런, 질감에 영향을 준다. 유화제는 HLB값(Hydrophilic – Lipophilic Balance)으로 분류되며 HLB값이 높을수록 친수성, 낮을수록 친유성을 나타낸다.

It stably combines water and fat present in gelato and promotes the interaction between fat and protein. It also plays a role in resisting temperature changes and overruns and affects texture. Emulsifiers are classified by HLB (Hydrophilic-Lipophilic Balance) value. The higher the HLB value, the more hydrophilic and more lipophilic when lower.

레시틴 (E322)

과거에는 레시틴이 풍부한 노른자가 젤라또의 천연 유화제로 사용되었다. 레시틴은 식물성 오일 견과류 대두에도 함유되어 있다. 대두 레시틴의 경우 특유의 강한 향을 가지고 있어 초콜릿이나 강한 젤라또 맛에만 어울리는 경향이 있다. 최대 0.3% 비율로 사용할 수 있다.

Lecithin (E322)

In the past, egg yolks were used as a natural emulsifier in gelato because they are rich in lecithin. Vegetable oils, nuts, and soybeans also include lecithin. As for soy lecithin, it's usually used only with chocolate or other intense gelato flavors because it has a unique, strong aroma. It can be used up to 0.3%.

모노 디 글리세라이드 (E471)

식물이나 동물성 지방에서 추출되는 모도 디 글리세라이드는 비건 젤라또에는 사용할 수 없는 유화제이다. 젤라또 안의 공기 포집을 개선시키며 젤라또를 보다 더 부드럽게 만들어준다. 뜨거운 온도에서 사용해야 하고 최대 0.5% 비율로 사용할 수 있다.

Mono- and Diglycerides (E471)

Mono- and Diglycerides is extracted from plant or animal fat and is an emulsifier that cannot be used in vegan gelato. It improves aeration inside the gelato and makes the gelato softer. It must be used at hot temperatures and can be used up to 0.5%.

자당 지방산 에스테르 (E473)

자당과 지방산의 결합에서 발생되는 유화제이다. 자당은 친수성을 지방산은 친유성을 가지고 있어 서로 잘 섞이지 않는 성분을 균일하게 결합시켜준다. 뜨거운 온도에서 사용해야 하고 최대 0.5% 비율로 사용할 수 있다. 모노 디 글리세라이드와 함께 사용하면 젤라또가 보다 더 부드러운 구조가 된다.

Sucrose Esters of Fatty Acids (E473)

It is an emulsifier derived from the combination of sucrose and fatty acids. Sucrose is hydrophilic, and fatty acids are lipophilic: so they uniformly bind ingredients that do not mix well with each other. It must be activated at hot temperatures and can be used up to 0.5%. Gelato will have a smoother structure when used with mono- and diglycerides.

Stabilizer (Base5) 복합안정제 (베이스5)

젤라또 원료회사에서 나오는 제품으로 증점제와 유화제가 혼합되어 있다. 직접 배합한 안정제 사용 시 정확한 용법(냉온수의 반응, 용량)을 모르고 사용하면 완성된 젤라또에 여러 가지 문제점이 발생한다. 복합안정제(베이스5)의 경우 원료회사에서 젤라또에 적용하기 이상적인 배합으로 조합을 한 제품이므로 0.3~0.5% 비율로 사용된다. 다만, 냉수에 작용하는 소르베또용 안정제인지, 온수에 작용하는 젤라또용 안정제인지는 구분해서 사용해야 한다. 원료회사마다 혼합된 증점제와 유화제가 다르기에 같은 용량을 사용하더라도 회사마다 완성되는 젤라또 질감이 다름을 느낄 수 있다. 따라서, 샘플 테스트를 통해 본인이 원하는 질감을 형성해주는 원료회사를 선택하는 것을 권장한다.

It is a product from the gelato company that contains a mixture of thickeners and emulsifiers. Using a self-formulated stabilizer without knowing the exact usage (reaction with hot and cold water, dosage) cause various problems in the completed gelato. For example, Base5 stabilizer is a product that is with the ideal formulation to be applied to gelato by the ingredient company. It can be used between 0.3~0.5%. However, be aware of whether it is for sorbetto that activates in cold water or gelato that activates in hot water. Since the thickener and emulsifier mixed in each ingredient company are different, the completed gelato can result in a different texture even if the same dosage is used. Therefore, selecting a company that provides the desired texture through sample testing is recommended.

Base50 베이스50

단일안정제 또는 복합안정제 사용 시 정량을 지켜 계량을 해야 원했던 질감의 젤라또가 완성된다. 처음 젤라또를 제조할 때 정량을 지키지 못하는 실수를 한다면 질감 변화가 크게 일어나므로(서걱거리는 식감, 찐득거리는 식감 등) 사용하기 좀 더 용이한 베이스50 제품을 사용할 수 있다. 젤라또 원료회사에서 나오는 제품으로 젤라또용 베이스50은 안정제, 포도당, 탈지분유가 혼합되어 있고 소르베또용 베이스50은 안정제, 포도당, 식이섬유가 혼합되어 있다. 젤라또용 베이스50은 우유 1L 기준 3~5% 사용되고, 소르베또용 베이스50은 과일의 성질에 따라 1.5~5% 사용된다.

When using a single stabilizer or pre-mixed stabilizer, such as Base5, make sure to respect the dosage to complete the gelato with the desired texture. If the dosage is not respected when making the gelato for the first time, the texture changes significantly (grainy or sticky); therefore, it's easier to use Base50 products. Base50 ranges are products made by gelato ingredient companies. Base50 for Gelato is a mixture of stabilizers, glucose, and skim milk powder; Base50 for Sorbetto is a mixture of stabilizers, glucose, and dietary fiber. The dosage of Base50 for Gelato is 3~5% to 1 liter of milk, and Base50 for Sorbetto is 1.5~5% depending on the nature of the fruit.

4 Fruits

과일

젤라떼리아에서 사용되는 과일은 생과일, 냉동 과일, 과일 퓌레, 과일 페이스트 사용에 따라 같은 과일이라도 다른 맛을 연출할 수 있다.

재료 마케팅에 중점을 두는 젤라떼리아에서는 생과일 사용을 선호하는데, 한 곳에서 과일을 공급받는다 하더라도 생물이기에 그날 그날의 맛, 향, 당도가 다를 수 있다. 좋은 품질의 생과일을 대량 구매하고 손질하여 급속 냉동한 후 냉동 보관해 사용하면 계절에 상관없이 과일 젤라또와 소르베또를 만들 수 있다.

365일 일정한 과일 맛을 만들고 싶은 젤라떼리아에서는 과일 퓌레나 과일 페이스트 사용을 선호한다. 생과일과는 다르게 가공식품이기에 일정한 맛을 연출할 수 있다. 과일 퓌레의 경우 100% 과일 함유 무가당 퓌레와, 80~90% 과일 함유 가당 퓌레로 나뉘며, 냉동 유통이기 때문에 사용 전에 완전히 해동한 후 사용해야 한다. 완전히 해동하지 않고 사용할 경우 얼음 결정체가 크게 형성되어 젤라또 식감을 저해하는 요소가 된다.

젤라또 원료회사의 과일 페이스트로 만들 경우 단독으로만 사용하기에는 인위적인 맛이 강해 과일 퓌레나 생과일을 같이 섞어 사용하는 것을 권장한다.

직접 담은 과일 청을 사용해서 만들 수도 있지만, 당의 비율이 높기 때문에 많은 양을 사용하기에는 어려움이 있다.

Fruits used in the gelateria, such as fresh or frozen fruits, fruit purée, and fruit paste, can produce different flavors even if they are the same fruit depending on the use.

A gelateria that focuses on marketing the ingredients prefers to use fresh fruits. Even if the fruits are supplied from one place, the taste, aroma, and sweetness may differ from day to day because they are fresh produce. You can make fruit gelato and sorbetto regardless of the season by buying good quality fresh fruits in bulk, processing them, then blast-freeze to store frozen.

A gelateria that wants to make consistent fruit flavor 365 days a year prefers to use fruit purée or fruit paste. Unlike fresh fruits, these can produce a consistent taste because they are processed foods. Fruit purée is divided into 100% fruit containing unsweetened purée and 80~90% fruit containing sweetened purée, and since it's frozen, it must be completely thawed before use. When used without completely thawing, large ice crystals may form, which can degrade the texture of gelato.

When using fruit paste from a gelato ingredient company, mixing with fruit purée or fresh fruit is recommended due to the strong artificial taste when used alone.

You can also use fruit preserves you made yourself, but it isn't easy to use in large amounts because of its high sugar content.

과일의 종류는 과실 구조에 따라 크게 인과류, 준인과류, 핵과류, 장과류, 각과류(견과류) 5가지루 구분된다. 수박이나 참외는 채소인 박과류에 속하며, 최근 우리나라가 아열대성 기후가 되면서 열대과일도 따로 분류되고 있다.

인과류 : 꽃턱이 발달하여 과육이 형성되는 과일이다.
종류 : 사과, 배 등

준인과류 : 씨방이 발달하여 과육이 형성되는 과일이다.
종류 : 감, 레몬, 라임, 자몽, 오렌지, 귤, 유자 등

핵과류 : 내과피가 단단한 핵을 이루고 중과피가 과육을 이루는 과일이다.
종류 : 복숭아, 자두, 살구, 체리 등

장과류 : 꽃턱이 두꺼운 주머니 모양으로 한 개 또는 여러 개의 과실이 달리는 과일이다.
종류 : 포도, 무화과, 키위, 딸기, 블루베리, 라즈베리 등

각과류(견과류) : 단단한 외피 속에 있는 과일로 섭취하는 부분은 씨앗에 해당된다.
종류 : 호두, 밤, 잣, 개암(헤이즐넛), 아몬드, 피칸, 피스타치오 등

박과류 : 덩굴성 식물에 열리는 과일이다.
종류 : 수박, 멜론, 참외 등

열대과일 : 열대지방에서만 자라는 과일이다.
종류 : 바나나, 파인애플, 망고, 패션프루트, 그라비올라, 리치, 망고스틴, 두리안 등

Fruit is classified into five major categories by the structure of the fruits; pome fruits, quasi-pome fruits, stone fruits, berry fruits, and shell fruits (nuts). Watermelon and melon belong to the cucurbits family, which is a vegetable. Tropical fruits are also separately classified as Korea has recently become a subtropical climate.

Pome fruits : Fruits in which the flower receptacle develops and forms the pulp
Types : Apple, pear, etc.

Quasi-pome fruit : Fruits in which the ovary develops and forms the flesh
Types : Persimmon, lemon, lime, grapefruit, orange, mandarin, yuja, etc.

Stone fruit : Fruits in which the endocarp forms a hard core and the mesocarp forms the pulp
Types : Peach, plum, apricot, cherry, etc.

Berry fruit : Fruits with one or several fruits in the shape of a pocket with a thick flower receptacle
Types : Grape, fig, kiwi, strawberry, blueberry, raspberry, etc.

Shell fruit (nuts) : Fruits in a hard outer shell. The part eaten is the seed.
Types : Walnut, chestnut, pine nut, hazelnut, almond, pecan, pistachio, etc.

Cucurbits family : Fruits on vines
Types : Watermelon, melon, Korean melon, etc.

Tropical fruits : Fruits that grow only in the tropics
Types : Banana, pineapple, mango, passion fruit, graviola, lychee, mangosteen, durian, etc.

5 Chocolate
초콜릿

초콜릿 젤라또는 전세계에서 가장 많이 판매되는 맛이다. 카카오빈 가공 과정에서 파생되는 카카오매스, 카카오 파우더, 다크초콜릿, 밀크초콜릿, 화이트초콜릿 등을 사용해서 다양한 초콜릿 맛을 만들 수 있다. 다크초콜릿 기준으로 같은 초콜릿 레시피라 하더라도 선택하는 브랜드, 또는 카카오빈 생산지에 따라 최종 젤라또 맛이 달라진다.

Chocolate gelato is the best-selling flavor in the world. Various chocolate flavors can be made using cacao mass, cacao powder, dark, milk, and white chocolate derived from the cacao bean processing operation. Even if the recipe is the same using dark chocolate, the gelato will taste different depending on the brand you choose or the place where the cacao beans are produced.

카카오매스

카카오빈을 발효하고 세척 건조 후 열풍으로 로스팅하여 외피를 분리한 뒤 카카오닙스를 만들고 콘칭하면 카카오매스가 만들어진다. 카카오 리큐어라고도 불리며, 50% 카카오버터를 함유한다.

카카오 파우더

카카오매스를 압축하여 카카오버터의 일부를 분리하면 카카오 파우더가 된다. 젤라떼리아에서 사용되는 카카오 파우더는 지방 22~24%를 선호한다. 카카오 파우더 색상은 카카오빈의 로스팅 정도와 알칼리화 과정에 따라 달라질 수 있다.

70% 다크초콜릿

70% 카카오매스에 30% 설탕을 혼합하여 만든다. 밀크초콜릿과 화이트초콜릿처럼 분유가 들어가지 않아 비건 초콜릿 젤라또를 만들 때 사용할 수 있다.

밀크초콜릿

카카오매스에 설탕, 분유를 혼합하여 만든다.

화이트초콜릿

카카오버터에 설탕, 분유를 혼합하여 만든다.

루비 초콜릿

2017년 처음으로 소개된 다크, 밀크, 화이트에 이은 4세대 초콜릿이다. 색소가 첨가되어 루비 색을 내는 것이 아닌 카카오빈 자체가 붉은 빛을 띠고 있고 단맛과 새콤한 맛이 나는 것이 특징이다.

빈투바

초콜릿 메이커가 카카오빈 선별부터 초콜릿 생산까지 이르는 모든 공정을 직접 관여하여 만드는 초콜릿이다. 초콜릿 메이커가 직접 카카오빈 생산지를 선택하기 때문에 지역별로 다양한 싱글 오리진 초콜릿 맛을 느낄 수 있다.

Cacao Mass

Cacao mass is made by fermenting, washing, drying, and roasting cacao beans with hot air, separating the skins, making cacao nibs, and conching them. It's also called cacao liqueur and contains 50% cacao butter.

Cacao Powder

Cacao powder is obtained by compressing the cacao mass and isolating some cacao butter. The cacao powder with 22~24% fat is preferred to use in gelaterias. Cacao powder's color can vary depending on the degree of roasting and alkalization of the beans.

70% Dark Chocolate

It is made by mixing 30% sugar with 70% cacao mass. It can be used to make vegan chocolate gelato since it does not contain milk powder, unlike milk and white chocolates.

Milk Chocolate

It is made by mixing cacao mass with sugar and milk powder.

White Chocolate

It is made by mixing cacao butter with sugar and milk powder.

Ruby Chocolate

It's the 4th generation of chocolate, following dark, milk, and white, introduced for the first time in 2017. It is characterized by the fact that the cacao bean itself has a reddish color and a sweet and sour taste.

Bean to Bar

It is chocolate made by a chocolate maker directly involved in all processes, from selecting the beans to chocolate production. Since the chocolate makers themselves choose the origins, they can offer a variety of single-origin chocolate flavors by region.

6 Infusion

인퓨전

향신료에는 풀, 열매, 나무껍질 등 다양한 종류가 있으며, 젤라또의 맛과 향을 돋우는 역할을 한다. 젤라또에서는 스파이스 계열과 허브류를 주로 사용하고, 재료 고유의 향을 고온과 저온에서 우려내 인퓨징 젤라또를 만든다.

Hot infusion 기법은 베이스나 우유에 인퓨전 재료를 넣어 65~75°C로 가열한 후 15~30분간 향을 더 우려낸 다음 인퓨전 재료만 걸러내고 젤라또 제조를 하는 방법이다.

Cold infusion 기법은 모든 재료를 4°C에 냉장 보관하여 12~24시간 향을 우려낸 후 인퓨전 재료만 걸러내고 젤라또 제조를 하는 방법이다.

스파이스 (HOT infusion)

주로 딱딱한 씨앗과 열매로 이루어져 있으며, 1Kg당 0.2~1%를 사용한다.

예) 시나몬, 팔각, 정향, 통카빈, 후추, 카다멈, 넛맥 등

허브 (COLD infusion)

부드러운 잎과 줄기를 가지고 있으며, 그 종류만도 100여가지가 넘는다. 1Kg당 1~3%를 사용한다.

예) 바질, 민트, 타임, 로즈마리, 고수, 딜, 쑥 등

찻잎&꽃잎 (HOT or COLD infusion)

다양한 찻잎과 꽃잎을 사용하여 젤라또에 우아한 풍미를 연출할 수 있다. 1Kg당 0.2~1%를 사용한다.

예) 녹차, 호지차, 홍차, 차이티, 장미, 라벤더, 히비스커스, 카모마일 등

바닐라 (HOT infusion)

인퓨전 기법을 활용하여 바닐라 젤라또를 만들 때 뿐만 아니라 기본적인 젤라또를 만들 때도 가장 많이 사용되는 인퓨전 재료가 바로 바닐라이다. 바닐라는 열대지방에서 자라나는 난초과 식물로 가늘고 긴 꼬투리(바닐라빈이라 불린다)에 풍부한 향을 가지고 있다. 바닐라는 자라나는 지역에 따라 분류되며 멕시코, 타히티, 마다가스카르, 인도네시아가 주 생산지이다. 1Kg 당 1~2개를 사용한다.

원두 (HOT or COLD infusion)

커피 젤라또를 만들기 위해 에스프레소를 추출하거나 커피 믹스를 첨가하여 만들 수도 있지만 원두를 직접 인퓨징해서 만들 수도 있다. 원두를 인퓨징하는 방법에 따라 브라운 커피 젤라또 또는 화이트 커피 젤라또가 된다.

예) HOT = 브라운 커피, COLD = 화이트 커피

원두 재배지와 로스팅 강도에 따라 같은 커피 젤라또 레시피여도 다른 풍미를 가진 젤라또가 완성된다. Kg 당 3~8%를 사용한다.

코코넛 (HOT infusion)

코코넛 밀크나 코코넛 퓌레를 사용하지 않고, 코코넛 플레이크를 인퓨징하는 기법을 사용하면 레시피의 계산의 복잡함 없이(코코넛은 지방을 많이 함유하고 있어 레시피 균형을 잘 맞춰주어야 한다.) 쉽고 맛있는 코코넛 젤라또를 만들 수 있다. 코코넛 플레이크를 오븐에 구워 사용하면 진한 코코넛 풍미를 느낄 수 있으며, 굽지 않고 바로 사용하면 깔끔한 코코넛 풍미를 느낄 수 있다. Kg 당 3~8%를 사용한다.

There are various types of spices, such as herbs, fruits, tree barks, etc., and they play a role in enhancing the taste and aroma of gelato. Spices and herbs are mainly used for gelato, and infused gelato is made by infusing the ingredients' unique flavors at high or low temperatures.

The hot infusion technique is a method where infusing ingredients are added to the base or milk, heated to 65~75°C, and then the fragrance is further brewed for 15~30 minutes, then the infused ingredients are sieved out to make gelato.

The cold infusion technique is a method of refrigerating all the ingredients at 4°C, brewing the flavor for 12~24 hours, then sieving out the infused ingredients to make gelato.

Spices (HOT infusion)

It mainly consists of hard seeds and fruits and uses 0.2~1% per 1 kilogram.

E.g. Cinnamon, star anise, cloves, Tonka bean, black pepper, cardamom, nutmeg, etc.

Herbs (COLD infusion)

Herbs have soft leaves and stems, with more than 100 kinds. Use 1~3% per 1 kilogram.

Tea Leaves & Flower Petals (HOT or COLD infusion)

A variety of tea leaves and petals can be used to give an elegant flavor to gelato. Use 0.2~1% per 1 kilogram.

E.g. Green tea, Hojicha, black tea, chai tea, roses, lavender, hibiscus, chamomile, etc.

Vanilla (HOT infusion)

Vanilla is the most used infusing ingredient not only when making vanilla gelato but also for basic gelato. Vanilla is an orchid plant that grows in the tropics and has a rich fragrance in long, thin pods (called vanilla beans). Vanilla is classified by the region where it grows, and Mexico, Madagascar, and Indonesia are the primary producers. Use 1~2 pods per 1 kilogram.

Coffee Beans (HOT or COLD infusion)

You can make coffee-flavored gelato by extracting espresso or adding instant coffee powders, but you can also infuse the beans yourself. Depending on the method of infusing the beans, it becomes brown coffee gelato or white coffee gelato.

E.g. The hot method will make brown gelato, and the cold method will make white gelato.

Even with the same coffee gelato recipe, you can make it into different flavors depending on the coffee bean plantation and roasting intensity. Use 3~8% per 1 kilogram.

Coconut (HOT infusion)

By infusing coconut flakes instead of coconut milk or purée, you can make easy and delicious coconut gelato without the complexity of recipe calculations. (Coconut contains a lot of fat, so you must balance the recipe well.) If you use coconut flakes roasted in the oven, you can taste rich coconut flavor, and if you use them as is without baking, you can get a refreshing coconut flavor. Use 3~8% per 1 kilogram.

7 Plant-based Drinks

식물성 음료

전 세계적으로 종교, 환경 문제, 동물의 존엄성, 건강 관리 등의 이슈로 비건 인구가 증가하고 있다. 우리나라는 현재 전체 인구의 4%인 약 200만 명의 채식주의자들이 있고, 채식주의자는 아니지만 MZ세대에서도 비건 음식과 비건 디저트에 대한 관심이 하나의 트렌드로 확산되고 있다.

채식주의자의 유형은 아래와 같이 세분화되어 있다. 유제품을 허용하는 락토, 락토오보, 페스코, 폴로테리언, 플렉시테리언은 유제품이 들어간 젤라또를 문제없이 섭취할 수 있지만, 유제품을 허용하지 않는 푸르테리언, 비건, 오보들을 위한 젤라또는 모든 재료가 식물성이어야 한다.

The vegan population is increasing worldwide due to issues such as religion, environmental issues, animal dignity, and health care. There are currently about 2 million vegetarians in Korea, which is about 4% of the total population, and even among non-vegans, interest in vegan food and desserts is spreading as a trend in the MZ generation*.

*** MZ generation** : Refers collectively to the Millennials and Generation Z.

The types of vegetarians are subdivided as below. Lacto, lacto-ovo, pesco, pollotarians, and flexitarians, who allow dairy products, can consume dairy-based gelato without any problem. However, gelato for fruitarians, vegans, and ovo-vegans who do not allow dairy products must be made with all plant-based ingredients.

채식주의자 유형 Types of Vegetarians								
푸르테리언 Fruitarian	O	X	X	X	X	X	X	X
비건 Vegan	O	O	X	X	X	X	X	X
락토 Lacto	O	O	O	X	X	X	X	X
락토 오보 Lacto-Ovo	O	O	O	O	X	X	X	X
오보 Ovo	O	O	X	O	X	X	X	X
페스코 Pesco	O	O	O	O	O	X	X	X
폴로테리언 Pollotarian	O	O	O	O	O	O	X	X
플렉시테리언 Flexitarian	O	O	O	O	O	O	O	X

따라서 젤라또에 가장 많이 사용되는 우유를 대체할 식물성 음료들이 사용되어야 한다. 우유는 브랜드가 바뀌어도 최종 젤라또 맛에 큰 영향을 미치지 않지만 식물성 음료는 무엇을 선택하는지에 따라 최종 젤라또 맛에 영향을 미치기 때문에 젤라띠에레가 원하는 최종 맛을 잘 고민한 후 선택해야 한다.

① A사 우유 + 바닐라 = 바닐라 젤라또
B사 우유 + 바닐라 = 바닐라 젤라또
② 두유 + 바닐라 = 두유 바닐라 젤라또
코코넛 밀크 + 바닐라 = 코코넛 바닐라 젤라또

우리나라에서 현재 쉽게 구입할 수 있는 식물성 음료는 두유, 아몬드 밀크, 코코넛 밀크, 오트 밀크, 라이스 밀크, 캐슈넛 밀크가 있다. 이 중 식물성 음료 자체의 맛이 가장 약한 오트 밀크와 라이스 밀크를 사용하면 대체적으로 젤라또의 주재료 맛을 살릴 수가 있다.

Therefore, plant-based drinks should be used instead of milk, which is most often used in gelato. However, even if the brand changes, milk does not significantly affect the gelato taste, but the plant-based drinks you choose will affect the final taste. Accordingly, it is necessary to choose after careful consideration about the final taste that the gelatiere wants.

① Company A milk + vanilla = Vanilla gelato
Company B milk + vanilla = Vanilla gelato
② Soy milk + vanilla = Soy vanilla gelato
Coconut milk + vanilla = Coconut vanilla gelato

Plant-based drinks that can be easily purchased domestically are; soy milk, almond milk, coconut milk, oat milk, rice milk, and cashew milk. Among them, using oat milk and rice milk, which gives the weakest taste, can generally preserve the taste of the main ingredient of gelato.

8 Pastes

페이스트

젤라띠에레의 성향과 젤라떼리아의 운영 방향성에 따라 원재료만을 사용하는 곳, 원료(페이스트)만을 사용하는 곳, 이 두 가지를 혼합해서 사용하는 곳으로 제각각 다르다. 원재료만을 사용한 젤라또가 무조건 맛있다고 할 수 없으며, 원료를 사용한 젤라또가 무조건 맛없다고 할 수 없는 이유는 젤라띠에레가 주어진 재료를 어떻게 해석해서 만드느냐에 따라 젤라또의 퀄리티가 달라지기 때문이다.

원재료를 사용하는 젤라띠에레는 재료가 가지고 있는 특성과 고형분을 명확하게 알아야 균형 잡힌 젤라또를 만들 수 있고, 원료를 사용하는 젤라띠에레는 원료의 용도를 정확하게 알아야 인위적이지 않고 목넘김이 좋은 젤라또를 만들 수 있다.

젤라또 원료 회사에서 판매되는 원료는 크게 페이스트, 베이스(안정제), 토핑류가 있다.

Depending on the propensity of the gelatiere and the direction of operating the gelateria, some may use only raw ingredients, others may use only processed ingredients (paste), and some may use both. Gelato using only raw ingredients cannot be considered delicious, and gelato using only processed ingredients cannot be tasteless. The reason is that the quality of gelato varies depending on how the gelatiere interprets and makes with the given ingredients.

The gelatiere using the raw ingredients must clearly know the characteristics and solids content of the ingredients to make a well-balanced gelato. The gelatiere using the processed ingredients must know precisely how to use the ingredients to make a gelato that tastes good and is not artificial.

The processed ingredients sold by the gelato ingredient company are largely divided into pastes, bases (stabilizers), and toppings.

페이스트

젤라또의 맛을 내기 위해 사용되는 페이스트는 견과류, 과일, 향료로 분류된다.

지방 함량이 높은 견과류 페이스트는 향이나 색소가 들어가지 않은 원재료 함량 100% 제품을 사용하는 것을 권장한다. 특히나 피스타치오나 헤이즐넛의 경우 이탈리아산이 가장 품질이 높은데 국내에서는 원재료 구입이 쉽지 않아 원료회사에서 판매하는 시칠리아산 피스타치오 100% 페이스트와 피에몬테산 헤이즐넛 100% 페이스트를 사용하면 깊이감이 상당히 풍부한 젤라또를 만들 수 있다.

과일 페이스트와 향료 페이스트는 당이 50% 이상 들어 있다. 따라서 생과일이나 원재료로만 만들었을 때 부족할 수 있는 향미를 페이스트 소량의 페이스트를 혼합해 보완해줄 수 있다. 과일과 향료 페이스트 선택에 있어서도 원재료 함량이 높은 제품을 선택하는 것을 권장한다.

페이스트 사용 용량은 각 제품마다 권장하는 첨가량이 표시되어 있다. 그 지침을 따라도 되고 기호도에 맞게 조절해서 사용하면 된다. 제품별 영양 정보 표기 사항을 보고 고형분 함량을 꼭 확인한 후 레시피를 작성해야 균형 잡힌 젤라또를 만들 수 있다.

Pastes

Pastes used to flavor gelato are classified as nuts, fruits, and spices.

For nut pastes with high-fat content, it is recommended to use products with 100% raw ingredients content without fragrance or coloring. Italian pistachios and hazelnuts are of the highest quality, but buying raw ingredients in Korea is challenging. Instead, using 100% Sicilian pistachio paste and 100% Piedmont hazelnut paste sold by the ingredient companies will help you make gelato with a fairly deep richness.

Fruit or spice pastes contain more than 50% sugar. Therefore, the flavor that may lack when using only fresh fruits or raw ingredients can be supplemented by mixing a small amount of paste. It is recommended to choose products with high raw ingredient content when selecting fruit or spice pastes.

As for the amount of paste to be used, the recommended amount to use is instructed on each product. You can follow the instruction or adjust to your preference. You must read the nutrition information label for each product and check the solids content before writing the recipe to make a well-balanced gelato.

토핑류

토핑류에는 젤라또를 추출하는 과정에서 혼합할 수 있는 다양한 소스, 식감을 줄 수 있는 쿠키, 당 절임 과일 등이 있다. 토핑류를 직접 만들어 사용하기 위해서는 기초적인 제과 지식이 있어야 하고 오븐이 필요한 경우가 있는데 제과 기술이 없거나 제조할 수 있는 환경이 없는 곳에서는 원료회사에서 나오는 토핑류의 활용도가 상당히 높다.

Toppings

Toppings include various sauces that can be mixed in the process of extracting gelato, cookies to give texture, and candied fruits. In order to make and use toppings yourself, you must have basic confectionery knowledge and sometimes an oven. For places with no confectionery technology or production environment, using the toppings from the ingredient companies is highly useful.

Part 4.

Theory for Writing Gelato Recipes

젤라또 레시피를 작성하기 위한 이론

Note

- 레시피를 작성하는 방법은 여러 가지가 있지만, 그 중 가장 쉽게 결과값을 찾을 수 있는 Carpigiani Gelato University의 방법으로 설명하였다.
- 고형분에서 소수점은 1자리까지 봐주고, 중량에서 소수점은 올리거나 버렸다.

- There are many ways to write a recipe, but Carpigiani Gelato University's method was explained, which is the easiest to find the result value.
- The decimal point was counted to one digit, and the decimal point was raised or dropped for the weight.

1 Writing and Understanding Gelato Recipes

젤라또 레시피 작성과 이해

젤라또에서 레시피는 정해진 답이 없다. 젤라띠에레마다, 젤라떼리아가 속한 환경마다, 같은 재료를 사용하더라도 추구하는 최종 젤라또의 맛과 질감에 따라 너무나도 다른 해석이 가능하기 때문이다. 젤라또의 답은 젤라띠에레가 직접 내리지만 일반적으로 생각하는 바른 길로 가기 위한 길잡이는 존재한다. 그것이 바로 레시피의 '균형'이다. 젤라또 레시피는 '맛' 하나에만 포커스를 맞춘다고 맛있는 레시피라고 할 수 없다. 젤라또가 서빙되는 온도는 영하이고, 그 영하의 온도에서도 단맛, 재료의 풍미, 질감, 입 안에서 느껴지는 차가움의 정도 등 여러 요소가 전체적으로 조화를 이루어야 완벽한 젤라또가 만들어진다.

젤라또 레시피를 작성하고 이해하기 위해서는 젤라또에 사용되는 재료 각각의 고형분과 수분의 함량을 알아야 한다. 예를 들면 물은 100% 수분이지만, 같은 액체인 우유의 경우 87.5%가 수분이고 나머지는 고형분이다. 그렇기 때문에 물은 부피와 중량이 동일하지만 (1,000ml = 1,000g), 우유는 부피와 중량이 같을 수 없다(1,000ml = 1,030g).

✦ 젤라또를 만들 때 우유가 차지하는 비율이 60% 이상이므로 레시피를 우유 1L 기준으로 잡으면 계량이 편리하다.

젤라또 레시피는 크게 두 가지 방향성으로 나뉜다.

주재료의 성질을 더 상세하게 파악하여 그 재료에 어울리게 만드는 '**싱글 레시피**'와 하나의 베이스를 만들어두고 여기에 다른 재료들을 첨가해 다양한 레시피로 파생시키는 '**베이스 레시피**'이다.

싱글 레시피의 경우 주재료의 맛을 더 돋보이게 표현할 수 있는 장점이 있지만 레시피마다 각각의 가열 공정이 필요하므로 제조 시간이 오래 걸린다는 단점이 있다. 베이스 레시피의 경우 베이스를 만들어 두고 여기에 각각의 주재료를 첨가하는 방식이므로 제조 시간을 단축시키는 장점이 있지만, 주재료를 첨가함에 따라 달라지는 고형분들의 균형을 다시 맞춰주어야 한다는 단점이 있다.

어떤 방법이 좋다고 혹은 나쁘다고 이야기할 수는 없다. 어떤 방식을 선택하는지는 젤라띠에레의 성향과 젤라떼리아 제조 환경, 선택하는 기계에 따라 달라질 수 있는 부분이다.

젤라또 레시피 작성의 가장 첫 번째 순서는 균형 잡힌 젤라또를 만들기 위해 사용하는 재료들의 고형분을 정확하게 파악하는 것이다.

There is no set answer to the recipe in gelato. It's because different interpretations are possible depending on the taste and texture of the final gelato pursued, even if the same ingredients are used by each gelatiere and each environment to which the gelateria belongs. The result of gelato is given directly by the gelatiere, but there are guides to follow to the right path that is generally thought of. That is the 'balance' of the recipe. A gelato recipe cannot be said to be a good recipe if it focuses only on 'taste.' The temperature at which gelato is served is sub-zero, and even at sub-zero temperature, perfect gelato is made only when various factors such as sweetness, flavor, texture, and degree of coldness in the mouth are harmonized as a whole.

In order to formulate and understand a gelato recipe, you need to know the solids and moisture content of each ingredient used in gelato. For example, water is 100% moisture, but as for milk, which is in the same liquid form, 87.5% is moisture, and the rest is solid content. Therefore, water has the same volume and weight (1,000 ml = 1,000 g), but milk cannot have the same volume and weight (1,000 ml = 1,030 g).

✦ When making gelato, milk accounts for more than 60%, so it is convenient to measure the recipe based on 1L of milk.

Gelato recipes can be written in two directions.

One way is a 'single recipe' that identifies the properties of the main ingredient in more detail and makes it suitable for the according ingredient. The other is a 'base recipe' that you would make a base and add other ingredients to it to derive various recipes.

As for the single recipe, it has the advantage of being able to express the taste of the main ingredient more prominently but has the disadvantage that it takes a long time to make because each process requires a separate heating process. The base recipe has the advantage of shortening the time to make because it's a method of making the base and adding different main ingredients thereto. But it has the disadvantage of rebalancing the solids that change according to the addition of the main ingredient.

I cannot say one is better than the other. Which method to choose is a part that can vary depending on the propensity of the gelatiere, the manufacturing environment, and the machine you choose.

The first step in creating a gelato recipe is to accurately determine the solid content of the ingredients to make a balanced gelato.

젤라또에 사용되는 재료들의 평균 고형분 함량 Average solids content of ingredients used in gelato

유제품 Dairy product	당 Sugar	지방 Fat	무지유 고형분 Non-fat milk solids	기타 고형분 Other solids	총 고형분 Total solids	수분 Water
일반 우유 Whole milk	-	3.5%	9%	-	12.5%	87.5%
저지방 우유 Low-fat milk	-	1.8%	9%	-	10.8%	89.2%
무지방 우유 Non-fat milk	-	0.2%	9%	-	9.2%	90.8%
생크림 (35% 지방) Cream (35% fat)	-	35%	5.8%	-	40.8%	59.2%
생크림 (38% 지방) Cream (38% fat)	-	38%	5.6%	-	43.6%	56.4%
전지분유 Whole milk powder	-	26%	71%	-	97%	3%
탈지분유 Skim milk powder	-	1%	95%	-	96%	4%
버터 Butter	-	84%	-	-	84%	16%
가당 연유 Sweetened condensed milk	43%	9%	24%	-	76%	24%
무가당 연유 Unsweetened condensed milk	-	8~9%	18~24%	-	26~33%	74~67%
요거트 Yogurt	-	4%	9.5%	-	13.5%	86.5%
리코타 Ricotta	-	13%	12%	-	25%	75%
마스카르포네 Mascarpone	-	47%	8.5%	-	55.5%	44.5%
크림치즈 Cream cheese	-	31%	10%	-	41%	59%

당류 Sugars	당 Sugar	지방 Fat	무지유 고형분 Non-fat milk solids	기타 고형분 Other solids	총 고형분 Total solids	수분 Water
설탕 Sugar	100%	-	-	-	100%	-
함수결정포도당 Dextrose	92%	-	-	-	92%	8%
물엿 Glucose syrup	80%	-	-	-	80%	20%
글루코스 시럽 파우더 Glucose syrup powder	96%	-	-	-	96%	4%
전화당 Inverted sugar	75%	-	-	-	75%	25%
꿀 Honey	80%	-	-	-	80%	20%
트레할로스 Trehalose	100%	-	-	-	100%	-
말토덱스트린 Maltodextrin	96%	-	-	-	96%	4%
과당 Fructose	100%	-	-	-	100%	-
메이플 시럽 Maple syrup	67%	-	-	-	67%	33%
이눌린 Inulin	-	-	-	-	94%	4%

원재료 및 가공 제품 Raw material & processed products	당 Sugar	지방 Fat	무지유 고형분 Non-fat milk solids	기타 고형분 Other solids	총 고형분 Total solids	수분 Water
달걀 Eggs	-	14%	-	11%	25%	75%
노른자 Egg yolk	-	30%	-	18%	48%	52%
흰자 Egg white	-	-	-	15%	15%	85%
카카오 파우더 (10~12%) Cocoa powder (10~12% fat)	-	11%	-	84%	95%	5%
카카오 파우더 (22~24%) Cocoa powder (22~24% fat)	-	23%	-	72%	95%	5%
카카오매스 Cocoa mass	-	55%	-	44%	99%	1%
화이트초콜릿 White chocolate	55%	20%	15%	10%	100%	-
밀크초콜릿 Milk chocolate	45%	36%	12.5%	6.5%	100%	-
다크초콜릿 (70%) Dark chocolate (70%)	30%	40%	-	30%	100%	-
땅콩 Peanut	3%	50%	-	44%	97%	3%
아몬드 Almond	4.5%	55%	-	40.5%	100%	-
잣 Pine nut	4%	50%	-	42%	96%	4%
호두 Walnut	4%	68%	-	28%	100%	-
헤이즐넛 페이스트 Hazelnut paste	-	65%	-	35%	100%	-
피스타치오 페이스트 Pistachio paste	-	55%	-	45%	100%	-
베이스50 크림 (젤라또용) Base50 Cream (for gelato) 베이스50 cc Base50 cc	40%	-	40%	14%	94%	6%
베이스50 과일 (소르베또용) Base50 Fruit (for sorbetto) 베이스50 ff Base50 ff	80%	-	-	14%	94%	6%
복합안정제 Stabilizer	-	-	-	100%	100%	-

위 표는 일반적으로 젤라또에 사용되는 재료들의 평균 고형분 함량이므로, 정확한 함량은 사용하는 재료에 표기된 영양정보 란을 확인하거나, 표기가 되어 있지 않은 재료의 경우 인터넷 검색을 통해 꼭 확인한 후 레시피를 작성해야 한다. 재료의 성질과 고형분 함량을 파악하는 것은 레시피를 통해 만들어지는 젤라또의 맛과 질감을 자유롭게 조절하기 위한 가장 중요한 과정이다.

The table above shows the average solid content of ingredients commonly used in gelato. Therefore, for exact figures, check the nutritional information table marked in the ingredients used, or if the information is not printed, make sure to check them through an internet search before writing a recipe. Knowing the nature and solid content of ingredients is the most important process to freely adjust the taste and texture of gelato made through the recipe.

두 번째로 중요한 것은 화이트 베이스로 만든 젤라또 또는 싱글 레시피로 만든 젤라또의 최종 고형분 함량을 설정하는 것이다.

The second important thing is to establish the final solid content of gelato made with the white base or from a single recipe.

참고 안정적인 젤라또를 완성하기 위한 고형분 수치 가이드

Guide to solids content to make stable gelato

구분 Type	당 Sugar	지방 Fat	무지유 고형분 Non-fat milk solids	기타 고형분 Other solids	총 고형분 Total solids
화이트 베이스 White base	14~19%	3~7%	7~12%	0.2~0.5%	32~36%
싱글, 젤라또 Single, gelato	18~24%	7~12%	7~12%	0.2%↗	36~46%

◆ 위 표에서 '싱글, 젤라또'라 함은 싱글 레시피 젤라또, 베이스로 만든 젤라또 레시피를 말한다. (즉, 쇼케이스에 들어가는 '완성된' 젤라또의 고형분 수치이다.)

◆ In the above table, 'single, gelato' refers to a single recipe gelato or a gelato recipe made with a base. (In other words, it is the solids content of 'completed' gelato that goes into the showcase.)

젤라또는 일반적으로 -10°C ~ -14°C의 온도에 진열되어 판매가 되는데, 이 온도에서 젤라또가 너무 단단하지도 묽지도 않은 적당한 질감을 가지게 하기 위해서는 레시피 고형분의 균형이 중요하다. 각 고형분마다 적정한 수치가 가이드로 정해져 있으며, 이 가이드 내에서 숫자를 조합하면 쇼케이스에 진열했을 때 크고 작은 문제점이 발생하지 않는 젤라또를 만들 수 있다. 그러나 레시피의 수치를 완벽하게 맞춘다고 하더라도 쇼케이스를 놓는 위치, 젤라떼리아의 환경에 따라 젤라또의 질감은 변화될 수 있으니 경험치도 아주 중요하다.

Gelato is generally displayed and sold between the temperature of -10~-14°C. At this temperature, the balance of the solids of the recipe is essential to ensure the gelato has the right texture, neither too hard nor runny. An appropriate value for each solid content is set as a guide, and by combining the numbers within this guide, you can make a gelato that does not cause problems, large or small, when displayed in a showcase. However, even if the numbers in the recipe are perfectly set, the texture of gelato can change depending on the location of the showcase and the environment of the gelateria; therefore, experience is also very important.

베이스 레시피는 크게 두 가지로 나뉜다. 유제품이 들어가는 '화이트 베이스'와 유제품과 노른자가 함께 들어가는 '옐로우 베이스'이다. 화이트 베이스는 깔끔한 맛이 나므로 유지방의 풍미를 강하게 표현하고 싶을 때 사용한다. 옐로우 베이스는 유지방의 풍미보다 더 진하고 깊이 있는 풍미로 표현하고 싶을 때 사용한다.

화이트 베이스는 젤라띠에레가 원하는 어떤 재료의 맛이든 그대로 연출할 수 있어 젤라떼리아에서 가장 많이 사용되는 기본 베이스이다.

The base recipes are divided mainly into two main categories. One is a 'white base' with dairy products, and the other is a 'yellow base' with dairy products and egg yolks. The white base is used to create a clean taste or a strong flavor of milk fat. The yellow base is used when you want a richer and deeper flavor than milk fat.

The white base is the most used basic base in gelateria as it can produce the taste of any ingredient desired by the gelatiere.

2 White Base

화이트 베이스

❶ 베이스50 크림으로 만드는 화이트 베이스

White base made with Base50 cream

화이트 베이스 White base

재료 Ingredients		중량 (g) Weight	당 (%) Sugar	지방 (%) Fat	무지유 고형분 (%) Non-fat milk solids	기타 고형분 (%) Other solids	총 고형분 (%) Total solids
우유 Milk				(3.5)	(9)	-	
생크림 Cream (38% fat)			-	(38)	(5.6)	-	
탈지분유 Skim milk powder			-	-	(96)	-	
설탕 Sugar			(100)	-	-	-	
물엿 Glucose syrup			(80)	-	-	-	
베이스50 크림 Base50 cream			(40)	-	(40)	(14)	
TOTAL	g						
	%						

1. 재료와 재료의 고형분 함량을 적는다.

1. Write down the ingredients and the solid contents of the ingredients.

화이트 베이스 White base

재료 Ingredients		중량 (g) Weight	당 (%) Sugar	지방 (%) Fat	무지유 고형분 (%) Non-fat milk solids	기타 고형분 (%) Other solids	총 고형분 (%) Total solids
우유 Milk			-	(3.5)	(9)	-	
생크림 Cream (38% fat)			-	(38)	(5.6)	-	
탈지분유 Skim milk powder			-	-	(96)	-	
설탕 Sugar			(100)	-	-	-	
물엿 Glucose syrup			(80)	-	-	-	
베이스50 크림 Base50 cream			(40)	-	(40)	(14)	
TOTAL	g	10,000	1,700	600	1,050	50	3,400
	%		17%	6%	10.5%	0.5%	34%

2. 만들고자 하는 총 합계 중량을 설정한다.

- 일반적으로 베이스 레시피는 1,000g 또는 10,000g 기준으로 계산할 수 있다. 표를 참고하여 각 고형분의 %를 가이드(75p) 내에서 설정한다.

2. Set the total sum weight you want to make.

- Generally, the base recipe can be calculated on a 1,000 g or 10,000 g basis. Refer to the table and set the percentage (%) of each solids within the guide (p.75).

베이스50 크림으로 만드는 화이트 베이스 작성 풀이를
설명한 영상입니다. 이 과정을 이해한다면
이 책의 모든 레시피를 작성할 수 있으니 꼭 영상을 시청해주세요.

화이트 베이스 White base

재료 Ingredients		중량 (g) Weight	당 (%) Sugar	지방 (%) Fat	무지유 고형분 (%) Non-fat milk solids	기타 고형분 (%) Other solids	총 고형분 (%) Total solids
우유 Milk				(3.5)	(9)	-	
생크림 Cream (38% fat)			-	(38)	(5.6)	-	
탈지분유 Skim milk powder			-	-	(96)	-	
설탕 Sugar			(100)	-	-	-	
물엿 Glucose syrup			(80)	-	-	-	
베이스50 크림 Base50 cream		**(2)** 357	(40) **(3)** 142.8	-	(40) **(4)** 142.8	(14) **(1)** 50	**(5)** 335.6
TOTAL	g	10,000	1,700	600	1,050	50	3,400
	%		17%	6%	10.5%	0.5%	34%

3. 레시피를 계산할 때는 밑에서부터 위로 당류 파트까지 올라가고, 유제품 파트에서는 위에서부터 밑으로 내려오며 계산한다.

→ 따라서 계산 순서는 ① 안정제(베이스50 크림), ② 당류(설탕, 물엿), ③ 유제품(우유, 생크림, 탈지분유)가 된다.

위 레시피의 경우 밑에서 위로 올라갈 때 처음 만나는 재료가 베이스50 크림이다.

어떤 고형분을 먼저 계산해야 하는지를 보자. 레시피를 통틀어 기타 고형분을 가진 재료는 베이스50 크림만 있으니 베이스50 크림 기타 고형분 함량부터 계산한다. 현재 필요한 기타 고형분 함량은 0.5%이고, 총 중량 10,000g 기준 50g이므로 50g **(1)** 이 그대로 베이스50 크림의 기타 고형분값이 된다.

베이스50 크림(당 40%, 무지유 고형분 40%, 기타 고형분 14%)의 나머지 당과 무지유 고형분값을 알기 위해서는 베이스50 크림의 중량부터 계산해야 한다.

중량 : 기타 고형분 50 ÷ 기타 고형분 14% = 357 **(2)**

당 : 중량 357 × 당 40% = 142.8 **(3)**

무지유 고형분 : 중량 357 × 무지유 고형분 40% = 142.8 **(4)**

총 고형분 : 당 142.8 + 무지유 고형분 142.8 + 기타 고형분 50 = 335.6 **(5)**

3. When calculating recipes, start from the bottom up to the sugars and from the top down to the dairy.

→ Therefore, the order of calculation is ① stabilizer (Base50 cream), ② Sugars (sugar, glucose syrup), ③ dairy products (milk, cream, skim milk powder).

In the case of the recipe above, the first ingredient you encounter going up is the Base50 cream.

Let's find out which solids should be calculated first. The only ingredient with other solids is Base50 cream, so calculate the Base50 cream's other solids content first. The current required other solids content is 0.5%, which is 50 grams based on the total weight of 10,000 grams; therefore, 50 grams **(1)** is the other solids value of Base50 cream.

To determine the remaining sugar and non-fat milk solids values for Base50 cream (40% sugar, 40% non-fat milk solids, 14% other solids), calculate the weight of the Base50 cream first.

Weight : Other solids 50 ÷ Other solids 14% = 357 **(2)**

Sugar : Weight 357 × Sugar 40% = 142.8 **(3)**

Non-fat milk solids : Weight 357 × Non-fat milk solids 40% = 142.8 **(4)**

Total solids : Sugar 142.8 + Non-fat milk solids 142.8 + Other solids 50 = 335.6 **(5)**

화이트 베이스 White base

재료 Ingredients		중량 (g) Weight	당 (%) Sugar	지방 (%) Fat	무지유 고형분 (%) Non-fat milk solids	기타 고형분 (%) Other solids	총 고형분 (%) Total solids
우유 Milk				(3.5)	(9)	-	
생크림 Cream (38% fat)			-	(38)	(5.6)	-	
탈지분유 Skim milk powder			-	-	(96)	-	
설탕 Sugar		**(2)** 1,246	(100) **(1)** 1,245.8	-	-	-	**(3)** 1,245.8
물엿 Glucose syrup		**(2)** 389	(80) **(1)** 311.4	-	-	-	**(3)** 311.4
베이스50 크림 Base50 cream		357	(40) 142.8	-	(40) 142.8	(14) 50	335.6
TOTAL	g	10,000	1,700	600	1,050	50	3,400
	%		17%	6%	10.5%	0.5%	34%

4. 다음으로 당의 비율을 나누고 여기에 따른 당량과 중량을 작성한다. 사용되는 당류와 그 비율을 어떻게 나누는지에 따라 입 안에서 느껴지는 감미도와 질감을 세세하게 조절할 수 있다.

설탕은 항상 메인 당의 역할을 해야 하므로 어떤 당류와 혼합하든 60% 이상을 차지해야 한다. 함수결정포도당과 물엿은 25% 이상, 말토덱스트린은 10% 이상 넣는 것은 권하지 않는다.

● 함수결정포도당이 많이 들어가게 되면 묽은 질감의 젤라또가 된다. 물엿과 말토덱스트린이 많이 들어가게 되면 찐득한 질감의 젤라또가 된다.

젤라또를 만들 때 보통 설탕 한 종류만 사용하지 않고 다른 당류들을 혼합하여 사용한다. 그 이유는 설탕만으로 만들었을 때 젤라또의 단맛이 너무 강하게 느껴지기 때문이다. 만약 다른 당류를 혼합하지 않고 설탕의 양만 낮춘다면 영하의 온도에서 젤라또가 단단하게 유지될 것이다. 그렇기 때문에 적당한 단맛과 젤라또의 질감을 유지하기 위해 설탕과 다른 당류들을 혼합하는 것이다.

이 레시피는 설탕 80%, 물엿 20%의 비율의 예시이다.

당의 총 무게 1,700에서 베이스50 크림의 당 무게 142.8g을 뺀 1,557.2에서의 비율을 계산한다.

설탕 (당 100%)

당 : 1,557.2 × 80% = 1,245.8

→ 설탕은 100% 고형분이므로 당량 **(1)**, 중량 **(2)**, 총 고형분량 **(3)**이 동일하다.

물엿 (당 80%, 수분 20%)

● 글루코스 파우더 사용 시 ⇒ 당 96%, 수분 4%

당 : 1,557.2 × 20% = 311.4 **(1)**

→ 물엿은 20% 수분을 포함하므로 중량은 20% 수분이 포함된 값을 계산한다.

중량 : 311.4 ÷ 당 80% = 389 **(2)**

총 고형분 : 311.4 **(3)**

4. Next, divide the sugar ratio and write down the weight and weight of sugar accordingly. Depending on the sugar used and how the ratio is divided, the sweetness and texture felt in the mouth can be controlled in detail.

Sugar should always serve as the main sweetness, so it should account for at least 60% of whatever kind of sugar is mixed with it. However, adding more than 25% of dextrose and glucose syrup and more than 10% of maltodextrin is not recommended.

- When a lot of dextrose is added, gelato becomes thin. If you use a lot of glucose syrup or maltodextrin, gelato will become sticky.

When making gelato, a mixture of different sugars is used rather than just one type of sugar. It's because gelato will taste too sweet when made only with sugar. If you lower the amount of sugar without mixing it with other sugars, the gelato will remain firm at sub-zero temperature. That is why sugar and other sugars are mixed to maintain the appropriate sweetness and texture of gelato.

This recipe is an example of a ratio of 80% sugar and 20% glucose syrup.

Calculate the ratio at 1,557.2 by subtracting the 142.8 grams of sugar of the Base50 cream from the total sugar weight of 1,700 grams.

Sugar (100% sugar)

Sugar : 1,577.2 × 80% = 1,245.8

→ Since sugar is 100% solids, sugar weight **(1)**, weight **(2)**, and total solids **(3)** are the same.

Glucose syrup (80% sugar, 20% water)

- When using glucose powder ⇒ 96% sugar, 4% water

Sugar : 1,557.2 × 20% = 311.4 **(1)**

→ Since glucose syrup contains 20% water, the weight is calculated as a value containing 20% water.

Weight : 311.4 ÷ Sugar 80% = 389 **(2)**

Total solids : 311.4 **(3)**

화이트 베이스 White base

재료 Ingredients		중량 (g) Weight	당 (%) Sugar	지방 (%) Fat	무지유 고형분 (%) Non-fat milk solids	기타 고형분 (%) Other solids	총 고형분 (%) Total solids
우유 Milk		**(1)** 6,750	-	(3.5) **(2)** 236.3	(9) **(3)** 607.5	-	**(4)** 843.8
생크림 Cream (38% fat)			-			-	
탈지분유 Skim milk powder			-	-		-	
설탕 Sugar		1,246	(100) 1,245.8	-	-	-	1,245.8
물엿 Glucose syrup		389	(80) 311.4	-	-	-	311.4
베이스50 크림 Base50 cream		357	(40) 142.8	-	(40) 142.8	(14) 50	335.6
TOTAL	g	10,000	1,700	600	1,050	50	3,400
	%		17%	6%	10.5%	0.5%	34%

5. 우유는 레시피에서 가장 많은 비율을 차지하는데 일반적으로 총 합계 중량에서 55~70%를 차지한다.

이 레시피는 67.5%의 비율로 계산하였다.

우유 (지방 3.5%, 무지유 고형분 9%)

중량 : 총 중량 10,000 × 67.5% = 6,750 **(1)**

지방 : 중량 6,750 × 지방 3.5 % = 236.3 **(2)**

무지유 고형분 : 중량 6,750 × 무지유 고형분 9% = 607.5 **(3)**

총 고형분 : 지방 236.3 + 무지유 고형분 607.5 = 843.8 **(4)**

5. Milk accounts for the largest portion of the recipe, typically 55~70% of the total weight.

The recipe was calculated at a ratio of 67.5%.

Milk (3.5% fat, 9% non-fat milk solids)

Weight : Total weight 10,000 × 67.5% = 6,750 **(1)**

Fat : Weight 6,750 × fat 3.5% = 236.3 **(2)**

Non-fat milk solids : Weight 6,750 × Non-fat milk solids 9% = 607.5 **(3)**

Total solids : Fat 236.6 + Non-fat milk solids 607.5 = 843.8 **(4)**

화이트 베이스 White base

재료 Ingredients		중량 (g) Weight	당 (%) Sugar	지방 (%) Fat	무지유 고형분 (%) Non-fat milk solids	기타 고형분 (%) Other solids	총 고형분 (%) Total solids
우유 Milk		6,750	-	(3.5) 236.3	(9) 607.5	-	843.8
생크림 Cream (38% fat)		**(2)** 957	-	(38) **(1)** 363.7	(5.6) **(3)** 53.6	-	**(4)** 417.3
탈지분유 Skim milk powder			-	-	(96)	-	
설탕 Sugar		1,246	(100) 1,245.8	-	-	-	1,245.8
물엿 Glucose syrup		389	(80) 311.4	-	-	-	311.4
베이스50 크림 Base50 cream		357	(40) 142.8	-	(40) 142.8	(14) 50	335.6
TOTAL	g	10,000	1,700	600	1,050	50	3,400
	%		17%	6%	10.5%	0.5%	34%

6. 생크림의 중량을 찾기 위해서는 지방의 값부터 계산한다.

생크림 (지방 38%, 무지유 고형분 5.6%)

지방 : 총 지방 600 – 우유 지방 236.3 = 363.7 **(1)**

중량 : 지방 363.7 ÷ 지방 38 % = 957 **(2)**

무지유 고형분 : 중량 957 × 무지유 고형분 5.6% = 53.6 **(3)**

총 고형분 : 지방 363.7 + 무지유 고형분 53.6 = 417.3 **(4)**

6. To find out the weight of the cream, start with the fat value.

Cream (38% fat, 5.6% non-fat milk solids)

Fat : Total fat 600 - Milk fat 236.3 = 363.7 **(1)**

Weight : Fat 363.7 ÷ Fat 38% = 957 **(2)**

Non-fat milk solids : Weight 957 × Non-fat milk solids 5.6% = 53.6 **(3)**

Total solids : Fat 363.7 + Non-fat milk solids 56.3 = 417.3 **(4)**

화이트 베이스 White base

재료 Ingredients		중량 (g) Weight	당 (%) Sugar	지방 (%) Fat	무지유 고형분 (%) Non-fat milk solids	기타 고형분 (%) Other solids	총 고형분 (%) Total solids
우유 Milk		6,750	-	(3.5) 236.3	(9) 607.5	-	843.8
생크림 Cream (38% fat)		957	-	(38) 363.7	(5.6) 53.6	-	417.3
탈지분유 Skim milk powder		**(2)** 256	-	-	(96) **(1)** 246.1	-	**(3)** 246.1
설탕 Sugar		1,246	(100) 1,245.8	-	-	-	1,245.8
물엿 Glucose syrup		389	(80) 311.4	-	-	-	311.4
베이스50 크림 Base50 cream		357	(40) 142.8	-	(40) 142.8	(14) 50	335.6
TOTAL	g	10,000	1,700	600	1,050	50	3,400
	%		17%	6%	10.5%	0.5%	34%

7. 탈지분유의 중량을 찾기 위해서는 무지유 고형분의 값부터 계산한다.

탈지분유 (지방 1%, 무지유 고형분 95%)

- 탈지분유의 경우 지방 1%와 무지유 고형분 95%로 이루어져 있지만, 여기에서는 계산의 편의상 지방값 1%를 생략하고 무지유 고형분 96%로 계산하였다.

무지유 고형분 : 총 무지유 고형분 1,050 –
우유 무지유 고형분 607.5 –
생크림 무지유 고형분 53.6 –
베이스50 크림 무지유 유고형분 142.8
= 246.1 **(1)**

중량 : 246.1 ÷ 무지유 고형분 96 % = 256 **(2)**

총 고형분 : 246.1 **(3)**

7. To find out the weight of the skim milk powder, start with the non-fat milk solids value.

Skim milk powder (1% fat, 95% non-fat milk solids)

- Skim milk powder consists of 1% fat and 95% non-fat milk solids. But for the convenience of calculation, the 1% fat value was omitted and calculated non-fat milk solids as 96%.

Non-fat milk solids : Total non-fat milk solids 1,050 -
Non-fat milk solids of milk 607.5 -
Non-fat milk solids of cream 53.6 -
Non-fat milk solids of Base50 cream
142.8 = 246.1 **(1)**

Weight : 246.1 ÷ Non-fat milk solids 96% = 256 **(2)**

Total solids : 246.1 **(3)**

8. 중량의 합계가 10,000이 나오는지 모든 값을 더해본다.

우유 6,750 + 생크림 957 + 탈지분유 256 + 설탕 1,246 + 물엿 389 + 베이스50 크림 357 = 9,955

합계 10,000g 기준 ± 500g 범위는 오차만큼 물이나 우유를 더 넣거나 빼고 만들어도 되고, 부족하거나 넘은 상태 그대로 만들어도 큰 문제가 되지 않는다. 그러나 500g 이상 오차 범위가 생긴 경우 우유의 비율을 조절해서 유제품만 다시 계산해준다.

지금처럼 45g이 부족할 때는 베이스 10,000g에서 45g의 물 또는 우유는 넣으나 마나 맛과 질감에 영향을 미치지 않는 수준이니 생략해도 된다.

합계 10,000g을 맞추고 싶은 경우 우유를 6,750 + 45 = 6,795g으로 넣어준다.

● 1,000g 기준 오차 범위 = ±50g

8. Add all the values to see if the sum of the weight is 10,000.

Milk 6,750 + Cream 957 + Skim milk powder 246 + Sugar 1,246 + Glucose syrup 389 + Base50 cream 357 = 9,955

In the range of ±500 grams based on a total of 10,000 grams, you can add or subtract more water or milk as much as the difference, and it is not a big problem even if you use it as is. However, if the difference is more than 500 grams, only dairy product is recalculated by adjusting the milk ratio.

If the difference is 45 grams, as above, adding that much water or milk from 10,000 grams of the base will not affect the taste and texture, so you can omit it.

If you want to make a total of 10,000 grams, add 6,750 + 45 = 6,795 grams of milk.

● ±50 grams error range based on 1,000 grams

9. 총 고형분의 합이 가로와 세로가 동일하게 3,400인지 확인한다. 가로와 세로의 합이 다르면 계산을 잘못한 것이고 잘못된 계산에 따라 중량이 잘못 설정되어 처음에 설정했던 고형분 비율로 제조할 수 없다.

총 고형분

가로 : 총 당 1,700 + 총 지방 600 + 총 무지유 고형분 1,050 + 총 기타 고형분 50 = 총 고형분 3,400

세로 : 우유 총 고형분 843.8 + 생크림 총 고형분 417.3 + 탈지분유 총 고형분 246.1 + 설탕 총 고형분 1,245.8 + 물엿 총 고형분 311.4 + 베이스50 크림 총 고형분 335.6 = 총 고형분 3,400

9. Make sure that the sum of the total solids is 3,400 equally on the column and row. If the sum of the column and row are different, the calculation is wrong. Hence, the weight is set incorrectly due to the incorrect calculation, so it cannot be used with the initially set solids content ratio.

Total solids

Row : Total sugar 1,700 + Total fat 600 + Total non-fat milk solids 1,050 + Total other solids 50 = Total solids 3,400

Column : Total solids for milk 843.8 + Total solids for cream 417.3 + Total solids for skim milk powder 246.1 + Total solids for sugar 1,245.8 + Total solids for glucose syrup 311.4 + Total solids for Base50 cream 335.6 = Total solids 3,400

❷ 복합안정제로 만드는 화이트 베이스

White base made with stabilizer

화이트 베이스 White base

재료 Ingredients		중량 (g) Weight	당 (%) Sugar	지방 (%) Fat	무지유 고형분 (%) Non-fat milk solids	기타 고형분 (%) Other solids	총 고형분 (%) Total solids
우유 Milk			-	(3.5)	(9)	-	
생크림 Cream (38% fat)			-	(38)	(5.6)	-	
탈지분유 Skim milk powder			-	-	(96)	-	
설탕 Sugar			(100)	-	-	-	
함수결정포도당 Dextrose			(92)	-	-	-	
물엿 Glucose syrup			(80)	-	-	-	
복합안정제 Stabilizer			-	-	-	(100)	
TOTAL	g						
	%						

1. 재료와 재료의 고형분 함량을 적는다.

1. Write down the ingredients and according solid contents.

화이트 베이스 White base

재료 Ingredients		중량 (g) Weight	당 (%) Sugar	지방 (%) Fat	무지유 고형분 (%) Non-fat milk solids	기타 고형분 (%) Other solids	총 고형분 (%) Total solids
우유 Milk			-	(3.5)	(9)	-	
생크림 Cream (38% fat)			-	(38)	(5.6)	-	
탈지분유 Skim milk powder			-	-	(96)	-	
설탕 Sugar			(100)	-	-	-	
함수결정포도당 Dextrose			(92)	-	-	-	
물엿 Glucose syrup			(80)	-	-	-	
복합안정제 Stabilizer			-	-	-	(100)	
TOTAL	g	10,000	1,700	600	1,050	50	3,400
	%		17%	6%	10.5%	0.5%	34%

2. 만들고자 하는 총 합계 중량을 설정한다.

- 일반적으로 베이스 레시피는 1,000g 또는 10,000g 기준으로 계산할 수 있다. 표를 참고하여 각 고형분의 %를 가이드 내에서 설정한다.

2. Set the total sum weight you want to make.

- In general, the base recipe can be calculated in a 1,000 g or 10,000 g basis. Refer to the table and set the percentage (%) of each solids within the guide.

화이트 베이스 White base

재료 Ingredients		중량 (g) Weight	당 (%) Sugar	지방 (%) Fat	무지유 고형분 (%) Non-fat milk solids	기타 고형분 (%) Other solids	총 고형분 (%) Total solids
우유 Milk			-	(3.5)	(9)	-	
생크림 Cream (38% fat)			-	(38)	(5.6)	-	
탈지분유 Skim milk powder			-	-	(96)	-	
설탕 Sugar			(100)	-	-	-	
함수결정포도당 Dextrose			(92)	-	-	-	
물엿 Glucose syrup			(80)	-	-	-	
복합안정제 Stabilizer		**(2)** 50	-	-	-	(100) **(1)** 50	**(3)** 50
TOTAL	g	10,000	1,700	600	1,050	50	3,400
	%		17%	6%	10.5%	0.5%	34%

3. 레시피를 계산할 때는 밑에서부터 위로 당류 파트까지 올라가고, 유제품 파트에서는 위에서부터 밑으로 내려오며 계산한다.

→ 따라서 계산 순서는 ① 복합안정제, ② 당류(설탕, 함수결정포도당, 물엿), ③ 유제품(우유, 생크림, 탈지분유)가 된다.

위 레시피의 경우 밑에서 위로 올라갈 때 처음 만나는 재료가 바로 복합안정제이다.

어떤 고형분을 먼저 계산해야 하는지를 알아보자. 레시피를 통틀어 기타 고형분을 가진 재료는 복합안정제만 있으니 복합안정제 기타 고형분 함량부터 계산한다. 현재 필요한 기타 고형분 함량은 0.5%이고, 총 중량 10,000g 기준에 50g이니 50g이 그대로 복합안정제 기타 고형분값이 된다.

복합안정제 (기타고형분 100%)

복합안정제는 100% 고형분이므로 기타 고형분 **(1)**, 중량 **(2)**, 총 고형분량 **(3)**이 동일하다.

3. When calculating recipes, start from the bottom up to the sugars and from the top down to the dairy.

→ Therefore, the order of calculation is ① stabilizer, ② Sugars (sugar, dextrose, glucose syrup), ③ dairy products (milk, cream, skim milk powder).

In the case of the recipe above, the first ingredient you encounter going up is the stabilizer.

Let's find out which solids should be calculated first. The only ingredient with other solids is the stabilizer, so calculate the stabilizer's other solids content. The current required other solids content is 0.5%, which is 50 grams based on the total weight of 10,000 grams; therefore, 50 grams is the other solids value of stabilizer.

Stabilizer (100% other solids)

Since the stabilizer is 100% solids, other solids **(1)**, weight **(2)**, and total solids **(3)** are the same.

화이트 베이스 White base

재료 Ingredients		중량 (g) Weight	당 (%) Sugar	지방 (%) Fat	무지유 고형분 (%) Non-fat milk solids	기타 고형분 (%) Other solids	총 고형분 (%) Total solids
우유 Milk				(3.5)	(9)	-	
생크림 Cream (38% fat)			-	(38)	(5.6)	-	
탈지분유 Skim milk powder			-	-	(96)	-	
설탕 Sugar		**(2)** 1,190	(100) **(1)** 1,190	-	-	-	**(3)** 1,190
함수결정포도당 Dextrose		**(2)** 185	(92) **(1)** 170	-	-	-	**(3)** 170
물엿 Glucose syrup		**(2)** 425	(80) **(1)** 340	-	-	-	**(3)** 340
복합안정제 Stabilizer		50	-	-	-	(100) 50	50
TOTAL	g	10,000	1,700	600	1,050	50	3,400
	%		17%	6%	10.5%	0.5%	34%

4. 다음으로 당의 비율을 나누고 여기에 따라 당량과 중량을 작성한다. 사용되는 당류와 그 비율을 어떻게 나누는지에 따라 입 안에서 느껴지는 감미도와 젤라또의 질감을 세세하게 조절할 수 있다.

설탕은 항상 메인 당의 역할을 해야 하므로 어떤 당류와 혼합하든 60% 이상 차지해야 한다. 함수결정포도당과 물엿은 25% 이상, 말토덱스트린은 10% 이상 넣는 것은 권하지 않는다.

- 함수결정포도당이 많이 들어가게 되면 묽은 질감의 젤라또가 된다. 물엿과 말토덱스트린이 많이 들어가게 되면 찐득한 질감의 젤라또가 된다.

젤라또를 만들 때 보통 설탕 한 종류만 사용하지 않고 다른 당류들을 혼합하여 사용한다. 그 이유는 설탕만으로 만들었을 때 젤라또의 단맛이 너무 강하게 느껴지기 때문이다. 만약 다른 당류를 혼합하지 않고 설탕의 양만 낮춘다면 영하의 온도에서 젤라또가 단단하게 유지될 것이다. 그렇기 때문에 적당한 단맛과 젤라또의 질감을 유지하기 위해 설탕과 다른 당류들을 혼합하는 것이다.

이 레시피는 설탕 70%, 함수결정포도당 10%, 물엿 20%의 비율의 예시이다.

당의 총 무게 1,700 에서 당류의 비율대로 계산한다.

설탕 (당 100%)

당 : 1,700 × 70% = 1,190

→ 설탕은 100% 고형분이므로 당량 **(1)**, 중량 **(2)**, 총 고형분량 **(3)**이 동일하다.

함수결정포도당 (당 92%, 수분 8%)

당 : 1,700 × 10% = 170 **(1)**

→ 함수결정포도당은 8% 수분을 포함하기에 중량은 8% 수분이 포함된 값을 계산한다.

중량 : 170 ÷ 당 92% = 185 **(2)**

총 고형분 : 170 **(3)**

물엿 (당 80%, 수분 20%)

- 글루코스 파우더 사용 시 (당 96%, 수분 4%)

당 : 1,700 × 20% = 340 **(1)**

→ 물엿은 20% 수분을 포함하기에 중량은 20% 수분이 포함된 값을 계산한다.

중량 : 340 ÷ 당 80% = 425 **(2)**

총 고형분 : 340 **(3)**

4. Next, divide the sugar ratio and write down the weight and weight of sugar accordingly. Depending on the sugar used and how the ratio is divided, the sweetness and texture felt in the mouth can be controlled in detail.

   Sugar should always serve as the main sweetness, so it should account for at least 60% of whatever kind of sugar is mixed with it. However, adding more than 25% of dextrose and glucose syrup and more than 10% of maltodextrin is not recommended.

   - When a lot of dextrose is added, gelato becomes thin. If you use a lot of glucose syrup or maltodextrin, gelato will become sticky.

   When making gelato, a mixture of different sugars is used rather than just one type of sugar. It's because gelato will taste too sweet when made only with sugar. If you lower the amount of sugar without mixing it with other sugars, the gelato will remain firm at sub-zero temperature. That is why sugar and other sugars are mixed to maintain the appropriate sweetness and texture of gelato.

   This recipe is an example of a ratio of 70% sugar, 10% dextrose, and 20% glucose syrup.

   Calculate according to the sugar ratio in the total weight of 1,700 g sugar.

   Sugar (100% sugar)

   Sugar : 1,700 × 70% = 1,190
   → Since sugar is 100% solids, sugar weight **(1)**, weight **(2)**, and total solids **(3)** are the same.

   Dextrose (92% sugar, 8% water)

   Sugar : 1,700 × 10% = 170 **(1)**
   → Since dextrose contains 8% water, the weight is calculated as a value containing 8% water.
   Weight : 170 ÷ Sugar 92% = 185 **(2)**
   Total solids : 170 **(3)**

   Glucose syrup (80% sugar, 20% water)

   - When using glucose powder → 96% sugar, 4% water

   Sugar : 1,700 × 20% = 340 **(1)**
   → Since glucose syrup contains 20% water, the weight is calculated as a value containing 20% water.
   Weight : 340 ÷ Sugar 80% = 425 **(2)**
   Total solids : 340 **(3)**

화이트 베이스 White base

재료 Ingredients		중량 (g) Weight	당 (%) Sugar	지방 (%) Fat	무지유 고형분 (%) Non-fat milk solids	기타 고형분 (%) Other solids	총 고형분 (%) Total solids
우유 Milk		**(1)** 6,750	-	(3.5) **(2)** 236.3	(9) **(3)** 607.5	-	**(4)** 843.8
생크림 Cream (38% fat)			-	(38)	(5.6)	-	
탈지분유 Skim milk powder			-	-	(96)	-	
설탕 Sugar		1,190	(100) 1,190	-	-	-	1,190
함수결정포도당 Dextrose		185	(92) 170	-	-	-	170
물엿 Glucose syrup		425	(80) 340	-	-	-	340
복합안정제 Stabilizer		50	-	-	-	(100) 50	50
TOTAL	g	10,000	1,700	600	1,050	50	3,400
	%		17%	6%	10.5%	0.5%	34%

5. 우유는 레시피에서 가장 많은 비율을 차지하는데 일반적으로 총 합계 중량에서 55~70%를 차지한다.

이 레시피는 67.5%의 비율로 계산하였다.

우유 (지방 3.5%, 무지유 고형분 9%)

중량 : 총 중량 10,000 × 67.5% = 6,750 **(1)**

지방 : 중량 6,750 × 지방 3.5 % = 236.3 **(2)**

무지유 고형분 : 중량 6,750 × 무지유 고형분 9% = 607.5 **(3)**

총 고형분 : 지방 236.3 + 무지유 고형분 607.5 = 843.8 **(4)**

5. Milk accounts for the largest portion of the recipe, typically 55~70% of the total weight.

The recipe was calculated at a ratio of 67.5%.

Milk (3.5% fat, 9% non-fat milk solids)

Weight : Total weight 10,000 × 67.5% = 6,750 **(1)**

Fat : Weight 6,750 × fat 3.5% = 236.3 **(2)**

Non-fat milk solids : Weight 6,750 × Non-fat milk solids 9% = 607.5 **(3)**

Total solids : Fat 236.6 + Non-fat milk solids 607.5 = 843.8 **(4)**

화이트 베이스 White base

재료 Ingredients		중량 (g) Weight	당 (%) Sugar	지방 (%) Fat	무지유 고형분 (%) Non-fat milk solids	기타 고형분 (%) Other solids	총 고형분 (%) Total solids
우유 Milk		6,750	-	(3.5) 236.3	(9) 607.5	-	843.8
생크림 Cream (38% fat)		**(2)** 957	-	(38) **(1)** 363.7	(5.6) **(3)** 53.6	-	**(4)** 417.3
탈지분유 Skim milk powder			-	-	(96)	-	
설탕 Sugar		1,190	(100) 1,190	-	-	-	1,190
함수결정포도당 Dextrose		185	(92) 170	-	-	-	170
물엿 Glucose syrup		425	(80) 340	-	-	-	340
복합안정제 Stabilizer		50	-	-	-	(100) 50	50
TOTAL	g	10,000	1,700	600	1,050	50	3,400
	%		17%	6%	10.5%	0.5%	34%

6. 생크림의 중량을 찾기 위해서는 지방값부터 계산한다.

생크림 (지방 38%, 무지유 고형분 5.6%)

지방 : 총 지방 600 – 우유 지방 236.3 = 363.7 **(1)**

중량 : 지방 363.7 ÷ 지방 38 % = 957 **(2)**

무지유 고형분 : 중량 957 × 무지유 고형분 5.6% = 53.6 **(3)**

총 고형분 : 지방 363.7 + 무지유 고형분 53.6 = 417.3 **(4)**

6. To find out the weight of the cream, start with the fat value.

Cream (38% fat, 5.6% non-fat milk solids)

Fat : Total fat 600 – Milk fat 236.3 = 363.7 **(1)**

Weight : Fat 363.7 ÷ Fat 38% = 957 **(2)**

Non-fat milk solids : Weight 957 × Non-fat milk solids 5.6% = 53.6 **(3)**

Total solids : Fat 363.7 + Non-fat milk solids 56.3 = 417.3 **(4)**

화이트 베이스 White base

재료 Ingredients		중량 (g) Weight	당 (%) Sugar	지방 (%) Fat	무지유 고형분 (%) Non-fat milk solids	기타 고형분 (%) Other solids	총 고형분 (%) Total solids
우유 Milk		6,750	-	(3.5) 236.3	(9) 607.5	-	843.8
생크림 Cream (38% fat)		957	-	(38) 363.7	(5.6) 53.6	-	417.3
탈지분유 Skim milk powder		**(2)** 405	-	-	(96) **(1)** 388.9	-	**(3)** 388.9
설탕 Sugar		1,190	(100) 1,190	-	-	-	1,190
함수결정포도당 Dextrose		185	(92) 170	-	-	-	170
물엿 Glucose syrup		425	(80) 340	-	-	-	340
복합안정제 Stabilizer		50	-	-	-	(100) 50	50
TOTAL	g	10,000	1,700	600	1,050	50	3,400
	%		17%	6%	10.5%	0.5%	34%

7. 탈지분유의 중량을 찾기 위해서는 무지유 고형분값부터 계산한다.

탈지분유 (지방 1%, 무지유 고형분 95%)

- 탈지분유의 경우 지방 1%와 무지유 고형분 95%로 이루어져 있지만, 여기에서는 계산의 편의상 지방값 1%를 생략하고 무지유 고형분 96%로 계산하였다.

무지유 고형분 : 총 무지유 고형분 1,050 –
우유 무지유 고형분 607.5 –
생크림 무지유 고형분 53.6 = 388.9 **(1)**

중량 : 388.9 ÷ 무지유 고형분 96 % = 405 **(2)**

총 고형분 : 388.9 **(3)**

7. To find out the weight of the skim milk powder, start with the non-fat milk solids value.

Skim milk powder (1% fat, 95% non-fat milk solids)

- Skim milk powder consists of 1% fat and 95% non-fat milk solids. But for the convenience of calculation, the 1% fat value was omitted and calculated non-fat milk solids as 96%.

Non-fat milk solids : Total non-fat milk solids 1,050 -
Non-fat milk solids of milk 607.5 -
Non-fat milk solids of cream 53.6
= 388.9 **(1)**

Weight : 388.9 ÷ Non-fat milk solids 96% = 405 **(2)**

Total solids : 388.9 **(3)**

8. 중량의 합계가 10,000이 나오는지 모든 값을 더해본다.

우유 6,750 + 생크림 957 + 탈지분유 405 + 설탕 1,190 +
함수결정포도당 185 + 물엿 425 + 복합안정제 50 = 9,962

합계 10,000g 기준 ± 500g 범위는 오차만큼 물이나 우유를 더 넣거나 빼고 만들어도 되고, 부족하거나 넘은 상태 그대로 만들어도 큰 문제가 되지 않는다. 그러나 500g 이상 오차 범위가 생긴 경우 우유의 비율을 조절해서 유제품만 다시 계산해준다.

지금처럼 38g이 부족할 때는 베이스 10,000g에서 38g의 물 또는 우유는 넣으나 마나 맛과 질감에 영향을 미치지 않는 수준이니 생략해도 된다.

합계 10,000g을 맞추고 싶은 경우 우유를 6,750 + 38
= 6,788g으로 넣어준다.

● 1,000g 기준 오차 범위 = ±50g

✦ 레시피를 작성하는 방법이 여러 가지 있지만, 그 중 가장 쉽게 결과값을 찾을 수 있는 Carpigiani Gelato University의 방법으로 설명하였다.

8. Add all the values to see if the sum of the weight is 10,000.

Milk 6,750 + Cream 957 + Skim milk powder 405 +
Sugar 1,190 + Dextrose 185 +
Glucose syrup 425 + Stabilizer 50 = 9,962

In the range of ±500 grams based on a total of 10,000 grams, you can add or subtract more water or milk as much as the difference, and it is not a big problem even if you use it as is. However, if the difference is more than 500 grams, only dairy product is recalculated by adjusting the milk ratio.

If the difference is 38 grams as above, adding that much water or milk from 10,000 grams of the base will not affect the taste and texture, so you can omit it.

If you want to make a total of 10,000 grams, add 6,750 + 38 = 6,788 grams of milk.

● ±50 grams error range based on 1,000 grams

✦ There are many ways to formulate a recipe, but I used the method from Carpigiani Gelato University, which is the easiest to calculate the final value, to explain.

9. 총 고형분의 합이 가로와 세로가 동일하게 3,400인지 확인한다. 가로와 세로의 합이 다르면 계산을 잘못한 것이고 잘못된 계산에 따라 중량이 잘못 설정되어 처음에 설정했던 고형분 비율로 제조할 수 없다.

총 고형분

가로 : 총 당 1,700 + 총 지방 600 + 총 무지유 고형분 1,050 +
총 기타 고형분 50 = 총 고형분 3,400

세로 : 우유 총 고형분 843.8 + 생크림 총 고형분 417.3 +
탈지분유 총 고형분 388.9 + 설탕 총 고형분 1,190 +
함수결정포도당 총 고형분 170 + 물엿 총 고형분 340 +
복합안정제 총 고형분 50 = 총 고형분 3,400

9. Make sure that the sum of the total solids is 3,400 equally on the column and row. If the sum of the column and row are different, the calculation is wrong. Hence, the weight is set incorrectly due to the incorrect calculation, so it cannot be used with the initially set solids content ratio.

Total solids

Row : Total sugar 1,700 + Total fat 600 +
Total non-fat milk solids 1,050 +
Total other solids 50 = Total solids 3,400

Column : Total solids for milk 843.8 +
Total solids for cream 417.3 +
Total solids for skim milk powder 388.9 +
Total solids for sugar 1,190 +
Total solids for dextrose 170 +
Total solids for glucose syrup 340 +
Total solids for stabilizer 50 = Total solids 3,400

❸ 화이트 베이스 제조

Making White Base

화이트 베이스(베이스50 크림) 우유 1L 기준
White Base (Base50 cream), based on 1 liter of milk

재료 Ingredients	우유 1L 기준 Based on 1L Milk	우유 10L 기준 (10배합) Based on 10L Milk (×10)
우유 Milk	1,030g (1pack)	10,300g (10packs)
생크림 Cream (38% fat)	145g	1,450g
탈지분유 Skim milk powder	39g	390g
설탕 Sugar	189g	1,890g
물엿 Glucose syrup	59g	590g
베이스50 크림 Base50 cream	50g	500g
TOTAL	1,512g	15,120g

화이트 베이스(안정제) 우유 1L 기준
White Base (Stabilizer), based on 1 liter of milk

재료 Ingredients	우유 1L 기준 Based on 1L Milk	우유 10L 기준 (10배합) Based on 10L Milk (×10)
우유 Milk	1,030g (1pack)	10,300g (10packs)
생크림 Cream (38% fat)	145g	1,450g
탈지분유 Skim milk powder	60g	610g
설탕 Sugar	180g	1,800g
함수결정포도당 Dextrose	26g	260g
물엿 Glucose syrup	64g	640g
복합안정제 Stabilizer	7g	70g
TOTAL	1,512g	15,120g

위의 배합은 앞의 화이트 베이스 레시피 계산에 따라 나온 중량을 우유 1L 기준으로 환산한 배합이다.

(우유 1L = 1,030g)

베이스 제조 시 가장 많이 사용되는 재료가 우유이기 때문에 우유 1L 기준으로 환산하면 필요한 양만큼 배수해서 제조하기 편리하다.

예를 들어 이 레시피를 10배합으로 제조할 경우, 우유 1L 기준 레시피에 곱하기 10을 해서 우유 10개를 계량하지 않고 바로 사용하면 된다.

책에 나오는 젤라또 레시피 중 화이트 베이스를 사용하는 레시피는 모두 이 레시피로 만들었다.

The above formula is a recipe that converts the weight from the previous white base recipe calculation based on 1 liter of milk.

(1L of milk = 1,030 g)

Since milk is the most used ingredient when making the base, it is convenient to multiply the required amount by converting it to 1 liter of milk.

For example, if you make this recipe by ten times, multiply the 1L milk-based recipe by ten and use it immediately without weighing ten packs of milk.

Among the gelato recipes in the book, the recipes using the white base were made with this recipe.

당 Sugar	지방 Fat	무지유 고형분 Non-fat milk solids	기타 고형분 Other solids	총 고형분 Total solids
17%	6%	10.5%	0.5%	34%

베이스 레시피 작성법에 따라 각자 원하는 베이스 고형분 비율에 맞춰 계산한 후 젤라또 레시피를 만들 경우, 최종 젤라또 고형분 비율은 책의 비율과 달라진다. 어떤 경우든 최종 젤라또 고형분 비율이 고형분 수치 가이드 안에 들어오면 쇼케이스 내에서 안정적인 젤라또를 유지할 수 있다.

The final gelato solids ratio will differ from the ratio in the book if you make a gelato recipe after calculating the base solids ratio you want according to the base recipe formula. Under any circumstances, you can maintain stable gelato in the showcase if the final solids percentage of gelato falls within the solids value guide.

참고 **안정적인 젤라또를 완성하기 위한 고형분 수치 가이드**
Guide to solids content to make stable gelato

구분 Type	당 Sugar	지방 Fat	무지유 고형분 Non-fat milk solids	기타 고형분 Other solids	총 고형분 Total solids
싱글, 젤라또 Single, gelato	18~24%	7~12%	7~12%	0.2% ↗	36~46%

✦ 위 표에서 '싱글, 젤라또'라 함은 싱글 레시피 젤라또, 베이스로 만든 젤라또 레시피를 말한다. (즉, 쇼케이스에 들어가는 '완성된' 젤라또의 고형분 수치이다.)

✦ In the above table, 'single, gelato' refers to a single recipe gelato or a gelato recipe made with a base. (In other words, it is the solids content of 'completed' gelato that goes into the showcase.)

제조 방법
Procedure

1. 우유를 40°C로 가열한다.
2. 레시피에 들어가는 분말류를 믹싱볼에 계량한 후 1에 넣고 가열한다.
3. 45°C가 되면 생크림을 넣는다.
4. 85°C로 가열한다.
5. 4°C로 냉각한다.
6. 12시간 숙성한 후 사용한다.

- 냉장고에서 최대 1주일간 보관하며 사용할 수 있다.
- 이 제조 방법은 살균기를 이용했을 때의 공정이다. 가열복합제조기를 이용할 때도 공정은 동일하지만, 우유와 분말류를 넣고 가열하기 전에 블렌더로 믹싱하면 가열복합제조기의 작은 주입구에 깔끔하게 넣을 수 있다.

1. Heat milk to 40°C.
2. Measure the dry ingredients for the recipe in a mixing bowl and add into 1 heat the mixture.
3. When it reaches 45°C, add cream.
4. Cook to 85°C.
5. Cool to 4°C.
6. Let rest for 12 hours to use.

- It can be stored in a refrigerator for up to one week.
- This method is a making process when using a pasteurizer. The process is the same when using the heating combined freezer, but if you mix milk and powders in a blender before heating, you can easily pour them into the freezer's small inlet.

❹ 화이트 베이스로 만드는 피오르 디 라떼 젤라또

Fior de Latte gelato made with White base

피오르 디 라떼 Fior de Latte

재료 Ingredients		중량 (g) Weight	당 (%) Sugar	지방 (%) Fat	무지유 고형분 (%) Non-fat milk solids	기타 고형분 (%) Other solids	총 고형분 (%) Total solids
화이트 베이스 White base		1,000	(17) 170	(6) 60	(10.5) 105	(0.5) 5	340
생크림(주재료) Cream (38% fat, main ingredient)		100	-	(38) 38	(5.6) 5.6	-	43.6
TOTAL	g	1,100	170	98	110.6	5	383.6
	%		**15.5%**	**8.9%**	10%	0.5%	34.9%
설탕(보완 재료) Sugar (supplement ingredient)		40	(100) 40	-	-	-	40
TOTAL	g	1,140	210	98	110.6	5	423.6
	%		18.4%	8.6%	9.7%	0.4%	37.1%

✦ 베이스를 사용하여 젤라또를 만들 때는 화이트 베이스 1,000g 기준으로 주재료와 보완 재료를 넣어준다.

✦ When making a gelato using the base, add the main and supplement ingredients based on 1,000 grams of white base.

화이트 베이스로 만드는 대표적인 젤라또는 순수한 우유 맛이 나는 '피오르 디 라떼Fior di Latte'이다.

'우유의 꽃'이라고 불리는 피오르 디 라떼의 주재료는 유지방의 풍미를 더 높이기 위해 생크림이 된다. 이미 균형이 잡힌 베이스에 주재료를 넣음으로 그 성질에 따라 균형이 무너지게 된다. 피오르 디 라떼의 경우 주재료로 선택된 생크림이 35~38% 지방을 함유하여 베이스의 지방값이 높아지고 당값은 떨어진다. 이 상태로 젤라또를 만들 경우 고형분 함량이 낮아 쇼케이스에서 젤라또의 질감이 딱딱해진다. 따라서 당을 더 첨가해 다시 균형을 맞춰주는 레시피 작업이 필요하다.

The classic gelato made with the white base is 'Fior di Latte,' which tastes like pure milk.

The main ingredient of Fior di Latte, which is called the 'flower of milk,' is cream to enhance the flavor of milk fat. By adding the main ingredient to an already balanced base, the balance collapses by nature. As for Fior di Latte, the cream selected as the main ingredient contains 35~38% of fat, so the fat value of the base increases and the sugar value decreases. If you make the gelato as is, because of the low solid content, the texture of the gelato will become hard in the showcase. Therefore, the recipe should be recalculated to rebalance by adding more sugar.

참고 **안정적인 젤라또를 완성하기 위한 고형분 수치 가이드**

Guide to solids content to make stable gelato

구분 Type	당 Sugar	지방 Fat	무지유 고형분 Non-fat milk solids	기타 고형분 Other solids	총 고형분 Total solids
화이트 베이스 White base	14~19%	3~7%	7~12%	0.2~0.5%	32~36%
싱글, 젤라또 Single, gelato	18~24%	7~12%	7~12%	0.2%↗	36~46%

✦ 위 표에서 '싱글, 젤라또'라 함은 싱글 레시피 젤라또, 베이스로 만든 젤라또 레시피를 말한다. (즉, 쇼케이스에 들어가는 '완성된' 젤라또의 고형분 수치이다.)

✦ In the above table, 'single, gelato' refers to a single recipe gelato or a gelato recipe made with a base. (In other words, it is the solids content of 'completed' gelato that goes into the showcase.)

최종 젤라또 레시피의 고형분 함량은 싱글, 젤라또 비율 안에서 숫자들이 조합되면 쇼케이스에 진열하기 이상적인 젤라또의 고형분 균형 값이 된다.

● 주재료가 지방이 아닌 당이 더 많을 경우 지방 재료로 보완해준다.

When the numbers are combined in the 'single, gelato' ratio, the solids content of the final gelato recipe becomes the solids balance value that is ideal for display in the showcase.

● If the main ingredient has more sugar than fat, supplement it with a fat ingredient.

제조 방법
Procedure

1. 비커에 숙성된 화이트 베이스를 계량한다.
2. 화이트 베이스에 생크림, 설탕을 넣고 핸드블렌더로 믹싱한다.
3. 제조기에 넣고 냉각 교반한다.
4. 바트 또는 카라피나에 추출한 후 급냉고에 넣는다.
5. 바로 판매할 경우 5분간 급속 냉동한 후 쇼케이스로 옮긴다. 바로 판매하지 않을 경우(여유분일 경우) 1시간 급속 냉동한 후 -18°C 이하에서 냉동 보관한다.

1. Measure the aged white base in a container.
2. Add cream and sugar to the base, and mix with a hand blender.
3. Pour into the machine and cold-churn the mixture.
4. Extract in a stainless-steel container or a carapina, then put in a blast freezer.
5. If you are selling it immediately, transfer it to the showcase after blast-freezing for 5 minutes. If not (if it is a spare), blast-freeze for one hour, then store frozen at -18°C or lower.

❺ 피오르 디 라떼로 응용하는 젤라또

Gelato made with Fior di Latte

쌀(리조), 과일 포셰(아마레나), 과일 시럽, 초콜릿(스트라차텔라), 밀크캐러멜 등을 부재료로 혼합하면 또 다른 맛이 완성된다.

You can make it into different flavors by mixing sub-ingredients such as rice (riso), fruit pochée (Amarena), fruit syrup, chocolate (stracciatella), milk caramel, etc.

가당 쌀
Sweetened rice

Ingredients		Quantity
쌀	Rice	100g
물	Water	750g
설탕	Sugar	130g
함수결정포도당	Dextrose	20g
TOTAL		1,000g

1. 쌀을 물에 불린다.
2. 냄비에 모든 재료를 넣고 쌀이 밥알이 살아 있는 죽이 되는 정도로 조리한다.
3. 냉장고에서 완전히 식힌 후 젤라또에 사용한다.

● 쌀을 조리할 때 원하는 향신료를 넣고 쌀에 향을 첨가할 수 있다.
예) 시나몬 스틱, 바닐라빈, 레몬제스트 등

1. Soak the rice sufficiently in water.
2. Put all the ingredients in a pot and cook until it resembles porridge but with firm enough rice.
3. Cool completely in a refrigerator, then use in the gelato.

● Cook with desired spices to make different flavors of riso by adding flavor to the rice. E.g., cinnamon stick, vanilla bean, lemon zest, etc.

과일 포셰
Fruit pochée

Ingredients		Quantity
과일	Fruits	600g
설탕	Sugar	300g
함수결정포도당	Dextrose	100g
TOTAL		1,000g

1. 믹싱볼에 과육 과일과 설탕, 함수결정포도당을 넣고 과육이 깨지지 않도록 버무린다.
2. 55°C로 예열된 오븐에 믹싱볼을 넣고 5~6시간 둔다.
3. 중간중간 믹싱볼에 가라앉은 당류들을 섞어 준다.
4. 완성된 포셰는 시럽을 걸러내고 과육을 젤라또에 사용한다.

● 포셰는 여러 가지 과일을 섞어 만들 수도 있고, 원하는 향신료를 과일과 함께 넣어 향을 첨가할 수 있다.

● 남은 시럽은 더 걸쭉하게 졸여 젤라또나 음료의 소스 또는 케이크 시트를 적시는 시럽으로 활용할 수 있다.

1. Put fruit pulps, sugar, and dextrose in a mixing bowl and gently toss without breaking the pulps.
2. Keep the mixing bowl in an oven preheated to 55°C for 5~6 hours.
3. Mix the sugars that have settled in the bowl occasionally.
4. Separate the syrup from the pulps of the completed pochée.

● You can make the pochée by mixing several kinds of fruits or adding flavor by using desired spices with the fruits.

● The remaining syrup can be reduced to a thicker consistency to use as a sauce for gelato or beverages or as a syrup to moisten cakes.

과일 시럽
Fruit syrup

Ingredients		Quantity
젤라틴	Gelatin	5g
물	Water	20g
과일 퓌레	Fruit purée	450g
설탕	Sugar	340g
물엿	Glucose syrup	185g
TOTAL		1,000g

1. 젤라틴을 물에 불린다.
2. 냄비에 과일 퓌레, 설탕, 물엿을 넣고 가열한다.
3. 2에 불린 젤라틴을 넣고 녹인 후 식힌다.
4. 시럽 용기에 담아 젤라또 토핑 소스로 사용한다.

1. Soak gelatin in water.
2. Heat fruit purée, sugar, and glucose syrup in a pot.
3. Mix 2 with the soaked gelatin and let cool.
4. Put in a syrup bottle and use as a topping sauce for gelato.

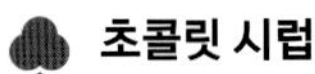

초콜릿 시럽
Chocolate syrup

Ingredients		Quantity
다크초콜릿	Dark chocolate	850(900)g
카카오버터	Cocoa butter	150(100)g
TOTAL		1,000g

1. 다크초콜릿과 카카오버터를 녹인다.
2. 1을 핸드블렌더로 잘 유화시킨다.
3. 시럽 용기에 담아 젤라또에 사용한다. 예) 스트라차텔라

● 초콜릿 시럽의 온도가 낮으면 굳어서 사용할 수 없으므로 젤라또에 사용하기 전 32℃ 정도로 맞춰 사용한다.

1. Melt dark chocolate and cocoa butter.
2. Emulsify 1 using a hand blender.
3. Put in a syrup bottle and use as a topping sauce for gelato. E.g., stracciatella

● If the temperature of the chocolate syrup is low, it hardens and cannot be used, so set it to about 32°C before use.

캐러멜 소스
Caramel sauce

Ingredients		Quantity
설탕	Sugar	450g
함수결정포도당	Dextrose	50g
생크림	Cream (38% fat)	400g
버터	Butter	100g
TOTAL		1,000g

1. 물기 없는 냄비에 설탕, 함수결정포도당을 계량한다.
2. 스패출러로 저어가며 가열해 녹인다.
3. 160~180°C에 도달하면 불을 끄고 생크림을 세 번 정도 천천히 나눠 넣어가며 저어준다.
4. 3에 버터를 넣어 녹인 후 식힌다.
5. 시럽 용기에 담아 젤라또 토핑 소스로 사용한다.

1. Measure sugar and dextrose in a dry pot.
2. Heat to melt while stirring with a spatula.
3. When it reaches 160~180°C, remove from heat and slowly add the cream in three portions while stirring.
4. Add butter to 3 to melt and let cool.
5. Put in a syrup bottle and use as a topping sauce for gelato.

3 Yellow Base

옐로우 베이스

옐로우 베이스의 대표적인 젤라또는 커스터드 크림 맛이 나는 '크레마(Crema)'이고, 유제품만 사용했을 때보다 깊이감이 풍부한 맛을 표현하고자 할 때 사용되는 베이스이다.

The representative gelato that uses the yellow base is 'Crema,' which tastes like a custard cream. It is used when the gelatiere wants to convey a taste that's rich in depth than when using only dairy products.

❶ 베이스50 크림으로 만드는 옐로우 베이스

Yellow base made with Base50 cream

옐로우 베이스 Yellow base

재료 Ingredients		중량 (g) Weight	당 (%) Sugar	지방 (%) Fat	무지유 고형분 (%) Non-fat milk solids	기타 고형분 (%) Other solids	총 고형분 (%) Total solids
우유 Milk			-	(3.5)	(9)	-	
생크림 Cream (38% fat)			-	(38)	(5.6)	-	
탈지분유 Skim milk powder			-	-	(96)	-	
설탕 Sugar			(100)	-	-	-	
물엿 Glucose syrup			(80)	-	-	-	
Base50 크림 Base50 cream			(40)	-	(40)	(14)	
노른자 Egg yolks			-	-	-	(18)	
TOTAL	g						
	%						

1. 재료와 재료의 고형분 함량을 적는다.

1. Write down the ingredients and according solid contents.

옐로우 베이스 레시피 작성법은 화이트 베이스와 동일하지만 주의할 점은 고형분 %를 설정할 때 기타 고형분과 총 고형분을 먼저 설정할 수가 없다는 것이다. Base50 크림 또는 복합안정제 이후에 주재료가 들어가는 경우(옐로우 베이스의 경우 노른자가 주재료가 된다.) 주재료에 따라 고형분이 기타 고형분을 포함하는 경우가 있어 항상 주재료 중량부터 설정하고 주재료 고형분 함량을 계산한 후 나머지 재료들은 화이트 베이스와 동일한 순서로 계산한다.

The yellow base recipe formulation is the same as the white base, but be aware that you cannot set other solids and totals solids first when setting the solids %. If the main ingredient is added after Base50 cream or stabilizer (for the yellow base, the main ingredient is egg yolks), solids may contain other solids depending on the main ingredient. Therefore, always set the main ingredient's weight first, and calculate the solids content of the main ingredient, then the rest of the ingredients are calculated in the same order as the white base.

옐로우 베이스 Yellow base

재료 Ingredients		중량 (g) Weight	당 (%) Sugar	지방 (%) Fat	무지유 고형분 (%) Non-fat milk solids	기타 고형분 (%) Other solids	총 고형분 (%) Total solids
우유 Milk			-	(3.5)	(9)	-	
생크림 Cream (38% fat)			-	(38)	(5.6)	-	
탈지분유 Skim milk powder			-	-	(96)	-	
설탕 Sugar			(100)	-	-	-	
물엿 Glucose syrup			(80)	-	-	-	
Base50 크림 Base50 cream			(40)	-	(40)	(14)	
노른자 Egg yolks			-	-	-	(18)	
TOTAL	g	10,000	1,950	750	800		
	%		19.5%	7.5%	8%		

2. 만들고자 하는 총 합계 중량을 설정한다.

- 일반적으로 베이스 레시피는 1,000g 또는 10,000g 기준으로 계산할 수 있다. 표를 참고하여 각 고형분의 %를 가이드 내에서 설정한다.

2. Set the total sum weight you want to make.

- Generally, the base recipe can be calculated on a 1,000 g or 10,000 g basis. Refer to the table and set the percentage (%) of each solids within the guide.

옐로우 베이스 Yellow base

재료 Ingredients		중량 (g) Weight	당 (%) Sugar	지방 (%) Fat	무지유 고형분 (%) Non-fat milk solids	기타 고형분 (%) Other solids	총 고형분 (%) Total solids
우유 Milk			-	(3.5)	(9)	-	
생크림 Cream (38% fat)			-	(38)	(5.6)	-	
탈지분유 Skim milk powder			-	-	(96)	-	
설탕 Sugar			(100)	-	-	-	
물엿 Glucose syrup			(80)	-	-	-	
Base50 크림 Base50 cream			(40)	-	(40)	(14)	
노른자 Egg yolks		600 **(1)**	-	(30) 180 **(2)**	-	(18) 108 **(3)**	288 **(4)**
TOTAL	g	10,000	1,950	750	800		
	%		19.5%	7.5%	8%		

3. 레시피를 계산할 때는 밑에서부터 위로 당류 파트까지 올라가고, 유제품 파트에서는 위에서부터 밑으로 내려오며 계산한다.

→ 따라서 계산 순서는 ① 노른자, ② 안정제(베이스50 크림), ③ 당류(설탕, 물엿), ④ 유제품(우유, 생크림, 탈지분유)가 된다.

이 레시피의 경우 밑에서 위로 올라갈 때 처음 만나는 재료가 노른자이다.

노른자의 첨가량은 5~10%이다. 노른자의 풍미를 얼마만큼 부여하고 싶은지에 따라 첨가량이 결정되며 이 레시피의 예시는 6%이다. 노른자 6%에 해당하는 고형분을 계산한다.

노른자 (지방 30%, 기타 고형분 18%)

중량 : 합계 10,000 × 노른자 6% = 600 **(1)**

지방 : 중량 600 × 지방 30% = 180 **(2)**

기타 고형분 : 중량 600 × 기타 고형분 18% = 108 **(3)**

총 고형분 : 지방 180 + 기타 고형분 108 = 288 **(4)**

3. When calculating recipes, start from the bottom up to the sugars and from the top down to the dairy.

→ Therefore, the order of calculation is ① egg yolks, ② stabilizer (Base50 cream), ③ Sugars (sugar, glucose syrup), ④ dairy products (milk, cream, skim milk powder).

In the case of the recipe above, the first ingredient you encounter going up is egg yolks.

The amount of egg yolks added is 5~10%. The amount depends on how much you want to impart the egg yolk flavor; the example in this recipe is 6%. Calculate the solids equivalent to 6% of egg yolks.

Egg yolks (30% fat, 18% other solids)

Weight : Total 10,000 × 6% Egg yolks = 600 **(1)**

Fat : Weight 600 × 30% Fat = 180 **(2)**

Other solids : Weight 600 × 18% Other solids = 108 **(3)**

Total solids : Fat 180 + Other solids 108 = 288 **(4)**

옐로우 베이스 Yellow base

재료 Ingredients		중량 (g) Weight	당 (%) Sugar	지방 (%) Fat	무지유 고형분 (%) Non-fat milk solids	기타 고형분 (%) Other solids	총 고형분 (%) Total solids
우유 Milk			-	(3.5)	(9)	-	
생크림 Cream (38% fat)			-	(38)	(5.6)	-	
탈지분유 Skim milk powder			-	-	(96)	-	
설탕 Sugar			(100)	-	-	-	
물엿 Glucose syrup			(80)	-	-	-	
Base50 크림 Base50 cream		350 (1)	(40) 140 (2)	-	(40) 140 (3)	(14) 49 (4)	329 (5)
노른자 Egg yolks		600	-	(30) 180	-	(18) 108	288
TOTAL	g	10,000	1,950	750	800		
	%		19.5%	7.5%	8%		

4. 다음으로 Base50 크림의 첨가량은 3~5%로 이 레시피의 예시는 3.5%이다. Base50 크림의 3.5%에 해당하는 고형분을 계산한다.

Base50 크림
(당 40%, 무지유 고형분 40%, 기타 고형분 14%)

중량 : 합계 10,000 × Base50 크림 3.5% = 350 **(1)**

당 : 중량 350 × 당 40% = 140 **(2)**

무지유 고형분 : 중량 350 × 무지유 고형분 40% = 140 **(3)**

기타 고형분 : 중량 350 × 기타 고형분 14% = 49 **(4)**

총 고형분 : 당 140 + 무지유 고형분 140 + 기타 고형분 49 = 329 **(5)**

4. Next, the amount of Base50 cream added is 3~5%; the example in this recipe is 3.5%. Calculate the solids equivalent to 3.5% of Base50 cream.

Base50 cream
(40% Sugar, 40% Non-fat milk solids, 14% Other solids)

Weight : Total 10,00 × 3.5% Base50 cream = 350 **(1)**

Sugar : Weight 350 × 40% Sugar = 140 **(2)**

Non-fat milk solids : Weight 350 × 40% Non-fat milk solids = 140 **(3)**

Other solids : Weight 350 × 14% Other solids = 49 **(4)**

Total solids : Sugar 140 + Non-fat milk solids 140 + Other solids 49 = 329 **(5)**

옐로우 베이스 Yellow base

재료 Ingredients		중량 (g) Weight	당 (%) Sugar	지방 (%) Fat	무지유 고형분 (%) Non-fat milk solids	기타 고형분 (%) Other solids	총 고형분 (%) Total solids
우유 Milk			-	(3.5)	(9)	-	
생크림 Cream (38% fat)			-	(38)	(5.6)	-	
탈지분유 Skim milk powder			-	-	(96)	-	
설탕 Sugar		1,448 (2)	(100) 1,448 (1)	-	-	-	1,448 (3)
물엿 Glucose syrup		453 (2)	(80) 362 (1)	-	-	-	362 (3)
Base50 크림 Base50 cream		350	(40) 140	-	(40) 140	(14) 49	329
노른자 Egg yolks		600	-	(30) 180	-	(18) 108	288
TOTAL	g	10,000	1,950	750	800		
	%		19.5%	7.5%	8%		

5. 다음으로 당의 비율을 나누고 여기에 따른 당량과 중량을 작성한다.

이 레시피는 설탕 80%, 물엿 20%의 비율의 예시이다.

당의 총 무게 1,950에서 Base50 크림의 당 무게 140을 뺀 1,810에서 비율을 계산한다.

설탕 (당 100%)

당 : 1,810 × 80% = 1,448

→ 설탕은 100% 고형분이므로 당량 (1), 중량 (2), 총 고형분량 (3)이 동일하다.

물엿 (당 80%, 수분 20%)

- 글루코스 파우더 사용 시 ⇒ 당 96%, 수분 4%

당 : 1,810 × 20% = 362 (1)

→ 물엿은 20% 수분을 포함하기에 중량은 20% 수분이 포함된 값을 계산한다.

중량 : 362 ÷ 당 80% = 452.5 (2)

총 고형분 : 362 (3)

5. Next, divide the ratio of sugar and write the sugar weight and weight accordingly.

The example in this recipe is 80% sugar and 20% glucose syrup.

Calculate the ratio from 1,810 by subtracting the Base50 cream's sugar weight of 140 from the total sugar weight of 1,950.

Sugar (100% sugar)

Sugar : 1,810 × 80% = 1,448

→ Sugar is 100% solids, so sugar weight (1), Weight (2), and Total solids (3) are the same.

Glucose syrup (80% Sugar, 20% Water)

- When using glucose powder → 96% Sugar, 4% Water

Sugar : 1,810 × 20% = 362 (1)

→ Glucose syrup contains 20% water, so calculate the weight value containing 20% water.

Weight : 362 ÷ 80% Sugar = 452.5 (2)

Total solids : 362 (3)

옐로우 베이스 Yellow base

재료 Ingredients		중량 (g) Weight	당 (%) Sugar	지방 (%) Fat	무지유 고형분 (%) Non-fat milk solids	기타 고형분 (%) Other solids	총 고형분 (%) Total solids
우유 Milk		6,100 (1)	-	(3.5) 213.5 (2)	(9) 549 (3)	-	762.5 (4)
생크림 Cream (38% fat)			-	(38)	(5.6)	-	
탈지분유 Skim milk powder			-	-	(96)	-	
설탕 Sugar		1,448	(100) 1,448	-	-	-	1,448
물엿 Glucose syrup		453	(80) 362	-	-	-	362
Base50 크림 Base50 cream		350	(40) 140	-	(40) 140	(14) 49	329
노른자 Egg yolks		600	-	(30) 180	-	(18) 108	288
TOTAL	g	10,000	1,950	750	800		
	%		19.5%	7.5%	8%		

6. 우유는 레시피에서 가장 많은 비율을 차지하는데 일반적으로 총 합계 중량에서 55~70%를 차지한다.

이 레시피는 61%의 비율로 계산하였다.

우유 (지방 3.5%, 무지유 고형분 9%)

중량 : 총 중량 10,000 × 61% = 6,100 **(1)**

지방 : 중량 6,100 × 지방 3.5 % = 213.5 **(2)**

무지유 고형분 : 중량 6,100 × 무지유 고형분 9% = 549 **(3)**

총 고형분 : 지방 213.5 + 무지유 고형분 549 = 762.5 **(4)**

6. Milk takes the largest portion of the recipe, typically 55~70% of the total weight. This recipe was calculated at 61%.

Milk (3.5% Fat, 9% Non-fat milk solids)

Weight : Total weight 10,000 × 61% = 6.100 **(1)**

Fat : Weight 6,100 × 3.5% Fat = 213.5 **(2)**

Non-fat milk solids : Weight 6,100 × 9% Non-fat milk solids = 549 **(3)**

Total solids : Fat 213.5 + Non-fat milk solids 549 = 762.5 **(4)**

옐로우 베이스 Yellow base

재료 Ingredients		중량 (g) Weight	당 (%) Sugar	지방 (%) Fat	무지유 고형분 (%) Non-fat milk solids	기타 고형분 (%) Other solids	총 고형분 (%) Total solids
우유 Milk		6,100	-	(3.5) 213.5	(9) 549	-	762.5
생크림 Cream (38% fat)		938 **(2)**	-	(38) 356.5 **(1)**	(5.6) 52.5 **(3)**	-	409 **(4)**
탈지분유 Skim milk powder			-	-	(96)	-	
설탕 Sugar		1,448	(100) 1,448	-	-	-	1,448
물엿 Glucose syrup		453	(80) 362	-	-	-	362
Base50 크림 Base50 cream		350	(40) 140	-	(40) 140	(14) 49	329
노른자 Egg yolks		600	-	(30) 180	-	(18) 108	288
TOTAL	g	10,000	1,950	750	800		
	%		19.5%	7.5%	8%		

7. 생크림의 중량을 찾기 위해서는 지방값부터 계산해준다.

생크림 (지방 38%, 무지유 고형분 5.6%)

지방 : 총 지방 750 – 우유 지방 213.5 – 노른자 지방 180
= 356.5 **(1)**

중량 : 지방 356.5 ÷ 지방 38 % = 938 **(2)**

무지유 고형분 : 중량 938 × 무지유 고형분 5.6% = 52.5 **(3)**

총 고형분 : 지방 356.5 + 무지유 고형분 52.5 = 409 **(4)**

7. To find the weight of the cream, calculate the fat value first.

Cream (38% Fat, 5.6% Non-fat milk solids)

Fat : Total fat 750 - Milk's fat 213.5 - Egg yolk's fat 180
= 356.5 **(1)**

Weight : Fat 356.5 ÷ 38% Fat = 938 **(2)**

Non-fat milk solids : Weight 938 ×
5.6% Non-fat milk solids = 52.5 **(3)**

Total solids : Fat 365.5 + Non-fat milk solids 52.5 = 409 **(4)**

옐로우 베이스 Yellow base

재료 Ingredients	중량 (g) Weight	당 (%) Sugar	지방 (%) Fat	무지유 고형분 (%) Non-fat milk solids	기타 고형분 (%) Other solids	총 고형분 (%) Total solids
우유 Milk	6,100	-	(3.5) 213.5	(9) 549	-	762.5
생크림 Cream (38% fat)	938	-	(38) 356.5	(5.6) 52.5	-	409
탈지분유 Skim milk powder	61 **(2)**	-	-	(96) 58.5 **(1)**	-	58.5 **(3)**
설탕 Sugar	1,448	(100) 1,448	-	-	-	1,448
물엿 Glucose syrup	453	(80) 362	-	-	-	362
Base50 크림 Base50 cream	350	(40) 140	-	(40) 140	(14) 49	329
노른자 Egg yolks	600	-	(30) 180	-	(18) 108	288
TOTAL g	10,000	1,950	750	800		
TOTAL %		19.5%	7.5%	8%		

8. 탈지분유의 중량을 찾기 위해서는 무지유 고형분값부터 계산해 준다.

탈지분유 (지방 1%, 무지유 고형분 95%)

- 탈지분유의 경우 지방 1%와 무지유 고형분 95%로 이루어져 있지만, 여기에서는 계산의 편의상 지방값 1%를 생략하고 무지유 고형분 96%로 계산하였다.

무지유 고형분 : 총 무지유 고형분 800 –
우유 무지유 고형분 549 –
생크림 무지유 고형분 52.5 –
Base50 크림 무지유 고형분 140 = 58.5 **(1)**

중량 : 58.5 ÷ 무지유 고형분 96 % = 60.9 **(2)**

총 고형분 : 58.5 **(3)**

8. To find the weight of skim milk powder, calculate the non-fat milk solids first.

Skim milk powder (1% Fat, 95% Non-fat milk solids)

- Skim milk powder consists of 1% fat and 95% non-fat milk solids. But for the convenience of calculation, the 1% fat value was omitted and calculated non-fat milk solids as 96%.

Non-fat milk solids : Total non-fat milk solids 800 -
Milk's non-fat milk solids 549 -
Cream's non-fat milk solids 52.5 -
Base50 cream's non-fat milk solids 140
= 58.5 **(1)**

Weight : 58.5 ÷ 96% Non-fat milk solids = 60.9 **(2)**

Total solids : 58.5 **(3)**

옐로우 베이스 Yellow base

재료 Ingredients		중량 (g) Weight	당 (%) Sugar	지방 (%) Fat	무지유 고형분 (%) Non-fat milk solids	기타 고형분 (%) Other solids	총 고형분 (%) Total solids
우유 Milk		6,100	-	(3.5) 213.5	(9) 549	-	762.5
생크림 Cream (38% fat)		938	-	(38) 356.5	(5.6) 52.5	-	409
탈지분유 Skim milk powder		61	-	-	(96) 58.5	-	58.5
설탕 Sugar		1,448	(100) 1,448	-	-	-	1,448
물엿 Glucose syrup		453	(80) 362	-	-	-	362
Base50 크림 Base50 cream		350	(40) 140	-	(40) 140	(14) 49	329
노른자 Egg yolks		600	-	(30) 180	-	(18) 108	288
TOTAL	g	10,000	1,950	750	800	157 (1)	3,657 (1)
	%		19.5%	7.5%	8%	1.57% (2)	36.57% (2)

9. 기타 고형분과 총 고형분을 계산한다.

기타 고형분 : Base50 크림 기타 고형분 49 +
노른자 기타 고형분 108 = 157 **(1)**

기타 고형분 비율 : 157 ÷ 10,000% = 1.57% **(2)**

총 고형분 : 우유 총 고형분 762.5 + 생크림 총 고형분 409 +
탈지분유 총 고형분 58.5 +
설탕 총 고형분 1,448 + 물엿 총 고형분 362 +
Base50 크림 총 고형분 329 +
노른자 총 고형분 288 = 3,657 **(1)**

총 고형분 비율 : 3,657 ÷ 10,000% = 36.57% **(2)**

9. Calculate other solids and total solids.

Other solids : Base50 cream's other solids 49 +
Egg yolk's other solids 108 = 157 **(1)**

Other solids ratio : 157 ÷ 10,000% = 1.57% **(2)**

Total solids : Milk's total solids 762.5 +
Cream's total solids 409 +
Skim milk powder's total solids 58.5 +
Sugar's total solids 1,448 +
Glucose syrup's total solids 362 +
Base50 cram's total solids 329 +
Egg yolk's total solids 288 = 3,657 **(1)**

Total solids ratio : 3,657 ÷ 10,000% = 36.57% **(2)**

10. 중량의 합계가 10,000이 나오는지 모든 값을 더해본다.

우유 6,100 + 생크림 938 + 탈지분유 61 + 설탕 1,448 +
물엿 453 + Base50 크림 350 + 노른자 600 = 9,950

합계 10,000g 기준 ±500g 범위는 오차만큼 물이나 우유를 더 넣거나 빼고 만들어도 되고, 부족하거나 넘은 상태 그대로 만들어도 큰 문제가 되지 않는다. 그러나 500g 이상 오차 범위가 생긴 겨우 우유의 비율을 조절해서 유제품만 다시 계산해준다.

지금처럼 50g이 부족할 때는 베이스 10,000g에서 50g의 물 또는 우유는 넣으나 마나 맛과 질감에 영향을 끼치지 않는 수준이니 생략해도 된다.

합계 10,000g을 맞추고 싶을 경우 우유를 6,100 + 50 = 6,150g 넣어준다.

● 1,000g 기준 오차 범위 = ±50g

10. Add all the values to check if the sum of the weight is 10,000.

Milk 6,100 + Cream 938 + Skim milk powder 61 +
Sugar 1,448 + Glucose syrup 453 + Base50 cream 350 +
Egg yolks 600 = 9,950

In the range of ±500 grams based on a total of 10,000 grams, you can add or subtract more water or milk as much as the difference, and it is not a big problem even if you use it as is. However, if the difference is more than 500 grams, only dairy product is recalculated by adjusting the milk ratio.

If the difference is 50 grams, as above, adding that much water or milk from 10,000 grams of the base will not affect the taste and texture, so you can omit it.

If you want to make a total of 10,000 grams, add 6,100 + 50 = 6,150 grams of milk.

● ±50 grams error range based on 1,000 grams

11. 총 고형분의 합이 가로와 세로가 동일하게 3,657인지 확인한다. 가로와 세로의 합이 다르면 계산을 잘못한 것이고 잘못된 계산에 따라 중량이 잘못 설정되어 처음에 설정했던 고형분 비율로 제조를 할 수 없다.

총 고형분

가로 : 당 1,950 + 지방 750 + 무지유 고형분 800 +
기타 고형분 157 = 3,657

세로 : 우유 총 고형분 762.5 + 생크림 총 고형분 409 +
탈지분유 총 고형분 58.5 + 설탕 총 고형분 1,448 +
물엿 총 고형분 362 + Base50 크림 총 고형분 329 +
노른자 총 고형분 288 = 3,657

11. Make sure that the sum of the total solids is 3,657 equally on the column and row. If the sum of the column and row are different, the calculation is wrong. Hence, the weight is set incorrectly due to the incorrect calculation, so it cannot be used with the initially set solids content ratio.

Total solids

Row : Sugar 1,950 + Fat 750 + Non-fat milk solids 800 +
Other solids 157 = 3,657

Column : Milk's total solids 762.5 +
Cream's total solids 409 +
Skim milk powder's total solids 58.5 +
Sugar's total solids 1,448 +
Glucose syrup's total solids 362 +
Base50 cream's total solids 329 +
Egg yolk's total solids 288 = 3,657

❷ 복합안정제로 만드는 옐로우 베이스

Yellow base made with stabilizer

옐로우 베이스 Yellow base

재료 Ingredients		중량 (g) Weight	당 (%) Sugar	지방 (%) Fat	무지유 고형분 (%) Non-fat milk solids	기타 고형분 (%) Other solids	총 고형분 (%) Total solids
우유 Milk			-	(3.5)	(9)	-	
생크림 Cream (38% fat)			-	(38)	(5.6)	-	
탈지분유 Skim milk powder			-	-	(96)	-	
설탕 Sugar			(100)	-	-	-	
함수결정포도당 Dextrose			(92)	-	-	-	
물엿 Glucose syrup			(80)	-	-	-	
복합안정제 Stabilizer			-	-	-	(100)	
노른자 Egg yolks			-	(30)	-	(18)	
TOTAL	g						
	%						

1. 재료와 재료의 고형분 함량을 적는다.

1. Write down the ingredients and according solids contents.

옐로우 베이스 Yellow base

재료 Ingredients		중량 (g) Weight	당 (%) Sugar	지방 (%) Fat	무지유 고형분 (%) Non-fat milk solids	기타 고형분 (%) Other solids	총 고형분 (%) Total solids
우유 Milk			-	(3.5)	(9)	-	
생크림 Cream (38% fat)			-	(38)	(5.6)	-	
탈지분유 Skim milk powder			-	-	(96)	-	
설탕 Sugar			(100)	-	-	-	
함수결정포도당 Dextrose			(92)	-	-	-	
물엿 Glucose syrup			(80)	-	-	-	
복합안정제 Stabilizer			-	-	-	(100)	
노른자 Egg yolks			-	(30)	-	(18)	
TOTAL	g	10,000	1,950	750	800		
	%		19.5%	7.5%	8%		

2. 만들고자 하는 총 합계 중량을 설정한다.

- 일반적으로 베이스 레시피는 1,000g 또는 10,000g 기준으로 계산할 수 있다. 표를 참고하여 각 고형분의 %를 가이드 내에서 설정한다.

2. Set the total sum weight you want to make.

- Generally, the base recipe can be calculated on a 1,000 g or 10,000 g basis. Refer to the table and set the percentage (%) of each solids within the guide.

옐로우 베이스 Yellow base

재료 Ingredients		중량 (g) Weight	당 (%) Sugar	지방 (%) Fat	무지유 고형분 (%) Non-fat milk solids	기타 고형분 (%) Other solids	총 고형분 (%) Total solids
우유 Milk			-	(3.5)	(9)	-	
생크림 Cream (38% fat)			-	(38)	(5.6)	-	
탈지분유 Skim milk powder			-	-	(96)	-	
설탕 Sugar			(100)	-	-	-	
함수결정포도당 Dextrose			(92)	-	-	-	
물엿 Glucose syrup			(80)	-	-	-	
복합안정제 Stabilizer			-	-	-	(100)	
노른자 Egg yolks		600 (1)	-	(30) 180 (2)	-	(18) 108 (3)	288 (4)
TOTAL	g	10,000	1,950	750	800		
	%		19.5%	7.5%	8%		

3. 레시피를 계산할 때는 밑에서부터 위로 당류 파트까지 올라가고, 유제품 파트에서는 위에서부터 밑으로 내려오며 계산한다.

→ 따라서 계산 순서는 ① 노른자, ② 복합안정제, ③ 당류(설탕, 함수결정포도당, 물엿), ④ 유제품(우유, 생크림, 탈지분유)가 된다.

이 레시피의 경우 밑에서 위로 올라갈 때 처음 만나는 재료가 노른자이다.

노른자의 첨가량은 5~10%이다. 노른자의 풍미를 얼만큼 부여하고 싶은지에 따라 첨가량이 결정되며 이 레시피의 예시는 6%이다. 노른자 6%에 해당하는 고형분을 계산한다.

노른자 (지방 30%, 기타 고형분 18%)

중량 : 합계 10,000 × 노른자 6% = 600 **(1)**

지방 : 중량 600 × 지방 30% = 180 **(2)**

기타 고형분 : : 중량 600 × 기타 고형분 18% = 108 **(3)**

총 고형분 : 지방 180 + 기타 고형분 108 = 288 **(4)**

3. When calculating recipes, start from the bottom up to the sugars and from the top down to the dairy.

→ Therefore, the order of calculation is ① egg yolks, ② stabilizer, ③ Sugars (sugar, dextrose, glucose syrup), ④ dairy products (milk, cream, skim milk powder).

In the case of the recipe above, the first ingredient you encounter going up is egg yolks.

The amount of egg yolks added is 5~10%. The amount depends on how much you want to impart the egg yolk flavor; the example in this recipe is 6%. Calculate the solids equivalent to 6% of egg yolks.

Egg yolks (30% fat, 18% other solids)

Weight : Total 10,000 × 6% Egg yolks = 600 **(1)**

Fat : Weight 600 × 30% Fat = 180 **(2)**

Other solids : Weight 600 × 18% Other solids = 108 **(3)**

Total solids : Fat 180 + Other solids 108 = 288 **(4)**

옐로우 베이스 Yellow base

재료 Ingredients		중량 (g) Weight	당 (%) Sugar	지방 (%) Fat	무지유 고형분 (%) Non-fat milk solids	기타 고형분 (%) Other solids	총 고형분 (%) Total solids
우유 Milk			-	(3.5)	(9)	-	
생크림 Cream (38% fat)			-	(38)	(5.6)	-	
탈지분유 Skim milk powder			-	-	(96)	-	
설탕 Sugar			(100)	-	-	-	
함수결정포도당 Dextrose			(92)	-	-	-	
물엿 Glucose syrup			(80)	-	-	-	
복합안정제 Stabilizer		35 **(1)**	-	-	-	(100) 35 **(2)**	35 **(3)**
노른자 Egg yolks		600	-	(30) 180	-	(18) 108	288
TOTAL	g	10,000	1,950	750	800		
	%		19.5%	7.5%	8%		

4. 다음으로 복합안정제의 첨가량은 0.3~0.5%로 이 레시피의 예시는 0.35%이다. 복합안정제 0.35%에 해당하는 고형분을 계산한다.

복합안정제 (기타 고형분 100%)

중량 : 합계 10,000 × 복합안정제 0.35% = 35 **(1)**

기타 고형분 : 중량 35 × 기타 고형분 100% = 35 **(2)**

총 고형분 : 35 **(3)**

4. Next, the amount of stabilizer added is 0.3~0.5%; the example in this recipe is 0.35%. Calculate the solids equivalent to 0.35% of stabilizer.

Stabilizer (100% Other solids)

Weight : Total 10,000 × 0.35% Stabilizer = 35 **(1)**

Other solids : Weight 35 × 100% Other solids = 35 **(2)**

Total solids : 35 **(3)**

옐로우 베이스 Yellow base

재료 Ingredients		중량 (g) Weight	당 (%) Sugar	지방 (%) Fat	무지유 고형분 (%) Non-fat milk solids	기타 고형분 (%) Other solids	총 고형분 (%) Total solids
우유 Milk			-	(3.5)	(9)	-	
생크림 Cream (38% fat)			-	(38)	(5.6)	-	
탈지분유 Skim milk powder			-	-	(96)	-	
설탕 Sugar		1,365 (2)	(100) 1,365 (1)	-	-	-	1,365 (3)
함수결정포도당 Dextrose		212 (2)	(92) 195 (1)	-	-	-	195 (3)
물엿 Glucose syrup		488 (2)	(80) 390 (1)	-	-	-	390 (3)
복합안정제 Stabilizer		35	-	-	-	(100) 35	35
노른자 Egg yolks		600	-	(30) 180	-	(18) 108	288
TOTAL	g	10,000	1,950	750	800		
	%		19.5%	7.5%	8%		

5. 다음으로 당의 비율을 나누고 여기에 따른 당량과 중량을 작성한다.

이 레시피는 설탕 70%, 함수결정포도당 10%, 물엿 20%의 비율의 예시이다.

당의 총 무게 1,950 에서 당류의 비율대로 계산한다.

설탕 (당 100%)

당 : 1,950 × 70% = 1,365

→ 설탕은 100% 고형분이므로 당량 (1), 중량 (2), 총 고형분량 (3)이 동일하다.

함수결정포도당 (당 92%, 수분 8%)

당 : 1,950 × 10% = 195 (1)

→ 함수결정포도당은 8% 수분을 포함하기에 중량은 8% 수분이 포함된 값을 계산한다.

중량 : 195 ÷ 당 92% = 211.9 (2)

총 고형분 : 211.9 (3)

5. Next, divide the ratio of sugar and write the sugar weight and weight accordingly.

The example in this recipe is 70% sugar, 10% dextrose, and 20% glucose syrup.

Calculate the ratio of sugars from the total sugar weight of 1,950.

Sugar (100% sugar)

Sugar : 1,950 × 70% = 1,365

→ Sugar is 100% solids, so sugar weight (1), Weight (2), and Total solids (3) are the same.

Dextrose (92% Sugar, 8% Water)

Sugar : 1,950 × 10% = 195 (1)

→ Dextrose contains 8% water, so calculate the weight value containing 8% water.

Weight : 195 ÷ 92% Sugar = 211.9 (2)

Total solids : 211.9 (3)

물엿 (당 80%, 수분 20%)

- 글루코스 파우더 사용 시 ⇒ 당 96%, 수분 4%

당 : 1,950 × 20% = 390 **(1)**

→ 물엿은 20% 수분을 포함하기에 중량은 20% 수분이 포함된 값을 계산한다.

중량 : 390 ÷ 당 80% = 487.5 **(2)**

총 고형분 : 390 **(3)**

Glucose syrup (80% Sugar, 20% Water)

- When using glucose powder → 96% Sugar, 4% Water

Sugar : 1,950 × 20% = 390 **(1)**

→ Glucose syrup contains 20% water, so calculate the weight value containing 20% water.

Weight : 390 ÷ 80% Sugar = 487.5 **(2)**

Total solids : 390 **(3)**

옐로우 베이스 Yellow base

재료 Ingredients		중량 (g) Weight	당 (%) Sugar	지방 (%) Fat	무지유 고형분 (%) Non-fat milk solids	기타 고형분 (%) Other solids	총 고형분 (%) Total solids
우유 Milk		6,100 (1)	-	(3.5) 213.5 (2)	(9) 549 (3)	-	762.5 (4)
생크림 Cream (38% fat)			-	(38)	(5.6)	-	
탈지분유 Skim milk powder			-	-	(96)	-	
설탕 Sugar		1,365	(100) 1,365	-	-	-	1,365
함수결정포도당 Dextrose		212	(92) 195	-	-	-	195
물엿 Glucose syrup		488	(80) 390	-	-	-	390
복합안정제 Stabilizer		35	-	-	-	(100) 35	35
노른자 Egg yolks		600	-	(30) 180	-	(18) 108	288
TOTAL	g	10,000	1,950	750	800		
	%		19.5%	7.5%	8%		

6. 우유는 레시피에서 가장 많은 비율을 차지하는데 일반적으로 총 합계 중량에서 55~70%를 차지한다.

이 레시피는 61%의 비율로 계산하였다.

우유 (지방 3.5%, 무지유 고형분 9%)

중량 : 총 중량 10,000 × 61% = 6,100 **(1)**

지방 : 중량 6,100 × 지방 3.5 % = 213.5 **(2)**

무지유 고형분 : 중량 6,100 × 무지유 고형분 9% = 549 **(3)**

총 고형분 : 지방 213.5 + 무지유 고형분 549 = 762.5 **(4)**

6. Milk takes the largest portion of the recipe, typically 55~70% of the total weight.

This recipe was calculated at 61%.

Milk (3.5% Fat, 9% Non-fat milk solids)

Weight : Total weight 10,000 × 61% = 6.100 **(1)**

Fat : Weight 6,100 × 3.5% Fat = 213.5 **(2)**

Non-fat milk solids : Weight 6,100 × 9% Non-fat milk solids = 549 **(3)**

Total solids : Fat 213.5 + Non-fat milk solids 549 = 762.5 **(4)**

옐로우 베이스 Yellow base

재료 Ingredients	중량 (g) Weight	당 (%) Sugar	지방 (%) Fat	무지유 고형분 (%) Non-fat milk solids	기타 고형분 (%) Other solids	총 고형분 (%) Total solids
우유 Milk	6,100	-	(3.5) 213.5	(9) 549	-	762.5
생크림 Cream (38% fat)	938 **(2)**	-	(38) 356.5 **(1)**	(5.6) 52.5 **(3)**	-	409 **(4)**
탈지분유 Skim milk powder		-	-	(96)	-	
설탕 Sugar	1,365	(100) 1,365	-	-	-	1,365
함수결정포도당 Dextrose	212	(92) 195	-	-	-	195
물엿 Glucose syrup	488	(80) 390	-	-	-	390
복합안정제 Stabilizer	35	-	-	-	(100) 35	35
노른자 Egg yolks	600	-	(30) 180	-	(18) 108	288
TOTAL g	10,000	1,950	750	800		
TOTAL %		19.5%	7.5%	8%		

7. 생크림의 중량을 찾기 위해서는 지방값부터 계산해준다.

생크림 (지방 38%, 무지유 고형분 5.6%)

지방 : 총 지방 750 – 우유 지방 213.5 – 노른자 지방 180 = 356.5 **(1)**

중량 : 지방 356.5 ÷ 지방 38 % = 938 **(2)**

무지유 고형분 : 중량 938 × 무지유 고형분 5.6% = 52.5 **(3)**

총 고형분 : 지방 356.5 + 무지유 고형분 52.5 = 409 **(4)**

7. To find the weight of the cream, calculate the fat value first.

Cream (38% Fat, 5.6% Non-fat milk solids)

Fat : Total fat 750 - Milk's fat 213.5 - Egg yolk's fat 180 = 356.5 **(1)**

Weight : Fat 356.5 ÷ 38% Fat = 938 **(2)**

Non-fat milk solids : Weight 938 × 5.6% Non-fat milk solids = 52.5 **(3)**

Total solids : Fat 365.5 + Non-fat milk solids 52.5 = 409 **(4)**

옐로우 베이스 Yellow base

재료 Ingredients		중량 (g) Weight	당 (%) Sugar	지방 (%) Fat	무지유 고형분 (%) Non-fat milk solids	기타 고형분 (%) Other solids	총 고형분 (%) Total solids
우유 Milk		6,100	-	(3.5) 213.5	(9) 549	-	762.5
생크림 Cream (38% fat)		938	-	(38) 356.5	(5.6) 52.5	-	409
탈지분유 Skim milk powder		207 (2)	-	-	(96) 198.5 (1)	-	198.5 (3)
설탕 Sugar		1,365	(100) 1,365	-	-	-	1,365
함수결정포도당 Dextrose		212	(92) 195	-	-	-	195
물엿 Glucose syrup		488	(80) 390	-	-	-	390
복합안정제 Stabilizer		35	-	-	-	(100) 35	35
노른자 Egg yolks		600	-	(30) 180	-	(18) 108	288
TOTAL	g	10,000	1,950	750	800		
	%		19.5%	7.5%	8%		

8. 탈지분유의 중량을 찾기 위해서는 무지유 고형분값부터 계산해 준다.

탈지분유 (지방 1%, 무지유 고형분 95%)

- 탈지분유의 경우 지방 1%와 무지유 고형분 95%로 이루어져 있지만, 여기에서는 계산의 편의상 지방값 1%를 생략하고 무지유 고형분 96%로 계산하였다.

무지유 고형분 : 총 무지유 고형분 800 –
우유 무지유 고형분 549 –
생크림 무지유 고형분 52.5 = 198.5 (1)

중량 : 198.5 ÷ 무지유 고형분 96 % = 206.7 (2)

총 고형분 : 198.5 (3)

8. To find the weight of skim milk powder, calculate the non-fat milk solids first.

Skim milk powder (1% Fat, 95% Non-fat milk solids)

- Skim milk powder consists of 1% fat and 95% non-fat milk solids. But for the convenience of calculation, the 1% fat value was omitted and calculated non-fat milk solids as 96%.

Non-fat milk solids : Total non-fat milk solids 800 -
Milk's non-fat milk solids 549 -
Cream's non-fat milk solids 52.5
= 198.5 (1)

Weight : 198.5 ÷ 96% Non-fat milk solids = 206.7 (2)

Total solids : 198.5 (3)

옐로우 베이스 Yellow base

재료 Ingredients		중량 (g) Weight	당 (%) Sugar	지방 (%) Fat	무지유 고형분 (%) Non-fat milk solids	기타 고형분 (%) Other solids	총 고형분 (%) Total solids
우유 Milk		6,100	-	(3.5) 213.5	(9) 549	-	762.5
생크림 Cream (38% fat)		938	-	(38) 356.5	(5.6) 52.5	-	409
탈지분유 Skim milk powder		207	-	-	(96) 198.5	-	198.5
설탕 Sugar		1,365	(100) 1,365	-	-	-	1,365
함수결정포도당 Dextrose		212	(92) 195	-	-	-	195
물엿 Glucose syrup		488	(80) 390	-	-	-	390
복합안정제 Stabilizer		35	-	-	-	(100) 35	35
노른자 Egg yolks		600	-	(30) 180	-	(18) 108	288
TOTAL	g	10,000	1,950	750	800	143 (1)	3,643 (1)
	%		19.5%	7.5%	8%	1.43% (2)	36.43% (2)

9. 기타 고형분과 총 고형분을 계산한다.

기타 고형분 : 복합안정제 기타 고형분 35 +
노른자 기타 고형분 108 = 143 **(1)**

기타 고형분 비율 : 143 ÷ 10,000% = 1.43% **(2)**

총 고형분 : 우유 총 고형분 762.5 + 생크림 총 고형분 409 +
탈지분유 총 고형분 198.5 + 설탕 총 고형분 1,365
+ 함수결정포도당 총 고형분 195 +
물엿 총 고형분 390 + 복합안정제 총 고형분 35 +
노른자 총 고형분 288 = 3,643 **(1)**

총 고형분 비율 : 3,643 ÷ 10,000% = 36.43% **(2)**

9. Calculate other solids and total solids.

Other solids : Stabilizer's other solids 35 +
Egg yolk's other solids 108 = 143 **(1)**

Other solids ratio : 143 ÷ 10,000% = 1.43% **(2)**

Total solids : Milk's total solids 762.5 +
Cream's total solids 409 +
Skim milk powder's total solids 198.5 +
Sugar's total solids 1,365 +
Dextrose's total solids 195 +
Glucose syrup's total solids 390 +
Stabilizer's total solids 35 +
Egg yolk's total solids 288 = 3,643 **(1)**

Total solids ratio : 3,643 ÷ 10,000% = 36.43% **(2)**

10. 중량의 합계가 10,000이 나오는지 모든 값을 더해본다.

우유 6,100 + 생크림 938 + 탈지분유 207 + 설탕 1,365 +
함수결정포도당 212 + 물엿 488 + 복합안정제 35 +
노른자 600 = 9,945

합계 10,000g 기준 ±500g 범위는 오차만큼 물이나 우유를 더 넣거나 빼고 만들어도 되고, 부족하거나 넘은 상태 그대로 만들어도 큰 문제가 되지 않는다. 그러나 500g 이상 오차 범위가 생긴 겨우 우유의 비율을 조절해서 유제품만 다시 계산해준다.

지금처럼 55g이 부족할 때는 베이스 10,000g에서 55g의 물 또는 우유는 넣으나 마나 맛과 질감에 영향을 끼치지 않으니 생략해도 된다.

합계 10,000g을 맞추고 싶을 경우 우유를 6,100 + 55 = 6,155g 넣어준다.

● 1,000g 기준 오차 범위 = ±50g

10. Add all the values to check if the sum of the weight is 10,000.

Milk 6,100 + Cream 938 + Skim milk powder 207 +
Sugar 1,365 + Dextrose 212 + Glucose syrup 488 +
Stabilizer 35 + Egg yolks 600 = 9,945

In the range of ±500 grams based on a total of 10,000 grams, you can add or subtract more water or milk as much as the difference, and it is not a big problem even if you use it as is. However, if the difference is more than 500 grams, only dairy product is recalculated by adjusting the milk ratio.

If the difference is 55 grams, as above, adding that much water or milk from 10,000 grams of the base will not affect the taste and texture, so you can omit it.

If you want to make a total of 10,000 grams, add 6,100 + 55 = 6,155 grams of milk.

● ±50 grams error range based on 1,000 grams

11. 총 고형분의 합이 가로와 세로가 동일하게 3,643인지 확인한다. 가로와 세로의 합이 다르면 계산을 잘못한 것이고 잘못된 계산에 따라 중량이 잘못 설정되어 처음에 설정했던 고형분 비율로 제조를 할 수 없다.

총 고형분

가로 : 당 1,950 + 지방 750 + 무지유 고형분 800 +
기타 고형분 143 = 3,643

세로 : 우유 총 고형분 762.5 + 생크림 총 고형분 409 +
탈지분유 총 고형분 198.5 + 설탕 총 고형분 1,365 +
함수결정포도당 총 고형분 195 + 물엿 총 고형분 390 +
복합안정제 총 고형분 35 + 노른자 총 고형분 288
= 3,643

11. Make sure that the sum of the total solids is 3,643 equally on the column and row. If the sum of the column and row are different, the calculation is wrong. Hence, the weight is set incorrectly due to the incorrect calculation, so it cannot be used with the initially set solids content ratio.

Total solids

Row : Sugar 1,950 + Fat 750 + Non-fat milk solids 800 +
Other solids 143 = 3,643

Column : Milk's total solids 762.5 +
Cream's total solids 409 +
Skim milk powder's total solids 198.5 +
Sugar's total solids 1,365 +
Dextrose's total solids 195 +
Glucose syrup's total solids 390 +
Stabilizer's total solids 35 +
Egg yolk's total solids 288 = 3,643

❸ 옐로우 베이스 제조

Making Yellow Base

옐로우 베이스(베이스50 크림) 우유 1L 기준
Yellow Base (Base50 Cream), based on 1 liter of milk

재료 Ingredients	우유 1L 기준 Based on 1L Milk	우유 10L 기준 (10배합) Based on 10L Milk (×10)
우유 Milk	1,030g (1pack)	10,300g (10packs)
생크림 Cream (38% fat)	157g	1,570g
탈지분유 Skim milk powder	10g	100g
설탕 Sugar	242g	2,420g
물엿 Glucose syrup	76g	760g
Base50 크림 Base50 cream	50g	500g
노른자 Egg yolks	100g	1,000g
TOTAL	1,665g	16,665g

옐로우 베이스(안정제) 우유 1L 기준
Yellow Base (Stabilizer), based on 1 liter of milk

재료 Ingredients	우유 1L 기준 Based on 1L Milk	우유 10L 기준 (10배합) Based on 10L Milk (×10)
우유 Milk	1,030g (1pack)	10,300g (10packs)
생크림 Cream (38% fat)	157g	1,570g
탈지분유 Skim milk powder	34g	340g
설탕 Sugar	228g	2,280g
함수결정포도당 Dextrose	35g	350g
물엿 Glucose syrup	81g	810g
복합안정제 Stabilizer	6g	60g
노른자 Egg yolks	100g	1,000g
TOTAL	1,671g	16,710g

위의 배합은 앞의 옐로우 베이스 레시피 계산에 따라 나온 중량을 우유 1L 기준으로 환산한 레시피이다.

(우유 1L = 1,030g)

베이스 제조 시 가장 많이 사용되는 재료가 우유 이기 때문에 우유 1L 기준으로 환산하면 필요한 양만큼 배수해서 제조하기 편리하다.

예를 들어 이 레시피를 10 배합으로 제조할 경우 우유 1L 기준 레시피에 곱하기 10을 해서 우유 10개를 계량하지 않고 바로 사용하면 된다.

책에 나오는 젤라또 레시피 중 옐로우 베이스를 사용하는 레시피는 이 레시피로 만들었다.

The above formula is a recipe that converts the weight from the previous yellow base recipe calculation based on 1 liter of milk.

(1L of milk = 1,030 g)

Since milk is the most used ingredient when making the base, it is convenient to multiply the required amount by converting it to 1 liter of milk.

For example, if you make this recipe by ten times, multiply the 1L milk-based recipe by ten and use it immediately without weighing ten packs of milk.

Among the gelato recipes in the book, the recipes using the yellow base were made with this recipe.

당 Sugar	지방 Fat	무지유 고형분 Non-fat milk solids	기타 고형분 Other solids	총 고형분 Total solids
19.5%	7.5%	8%	1.57%	36.57%

베이스 레시피 작성법에 따라 각자 원하는 베이스 고형분 비율에 맞춰 계산 후 젤라또 레시피를 만들 경우 최종 젤라또 고형분 비율은 책의 비율과 달라진다. 어떤 경우든 최종 젤라또 고형분 비율이 고형분 수치 가이드 안에 들어오면 쇼케이스 내에서 안정적인 젤라또를 유지할 수 있다.

The final gelato solids ratio will differ from the ratio in the book if you make a gelato recipe after calculating the base solids ratio you want according to the base recipe formula. Under any circumstances, you can maintain stable gelato in the showcase if the final solids percentage of gelato falls within the solids value guide.

참고 **안정적인 젤라또를 완성하기 위한 고형분 수치 가이드**
Guide to solids content to make stable gelato

구분 Type	당 Sugar	지방 Fat	무지유 고형분 Non-fat milk solids	기타 고형분 Other solids	총 고형분 Total solids
싱글, 젤라또 Single, gelato	18~24%	7~12%	7~12%	0.2%↗	36~46%

제조 방법
Procedure

1. 우유를 계량한다.
2. 노른자를 계량한다.
3. 레시피에 들어가는 분말류를 계량한다.
4. **2**와 **3**을 **1**에 넣고 핸드 블렌더로 믹싱한다.
5. 가열복합제조기 또는 냄비에 **4**를 넣는다.
6. 45°C가 되면 생크림을 넣고 65°C로 가열한 후 4°C로 냉각한다.
7. 12시간 숙성한 후 사용한다.

- 냉장고에서 최대 1주일간 보관하며 사용할 수 있다.
- 이 제조 방법은 가열복합제조기(또는 인덕션과 냄비)를 이용했을 때의 공정이다. 살균기를 이용할 때도 공정은 동일하지만, 가열복합제조기처럼 입구가 좁지 않으므로 핸드 블렌더로 믹싱하는 과정을 생략해도 된다.

1. Measure milk.
2. Measure egg yolks.
3. Measure the powders used in the recipe.
4. Add **2** and **3** into **1** and mix using a hand blender.
5. Put **4** in a heating combined freezer or a pot.
6. When it reaches 45°C, add cream and heat until 65°C, then cool to 4°C.
7. Age 12 hours to use.

- It can be stored in a refrigerator for up to one week.
- This method is a making process when using a heating combine freezer (or an induction and a pot). The process is the same when using a pasteurizer, but you can omit to mix with a hand blender because the inlet is not as narrow as in the heating combined freezer.

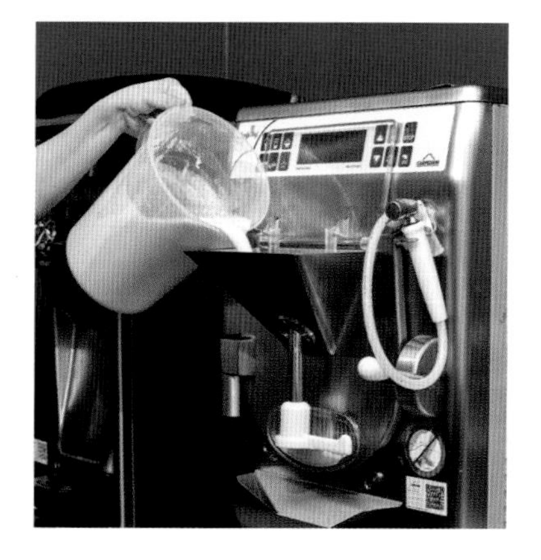

❹ 옐로우 베이스로 만드는 크레마 젤라또

Crema gelato made with Yellow base

크레마 Crema

재료 Ingredients		중량 (g) Weight	당 (%) Sugar	지방 (%) Fat	무지유 고형분 (%) Non-fat milk solids	기타 고형분 (%) Other solids	총 고형분 (%) Total solids
옐로우 베이스 Yellow base		1,000	(19.5) 195	(7.5) 75	(8) 80	(1.5) 15	365
생크림 Cream (38% fat)		80	-	(38) 30.4	(5.6) 4.5	-	34.9
레몬 제스트 Lemon zest		1개	-	-	-	-	-
TOTAL	g	1,080	195	105.4	84.5	15	399.8
	%		18%	9.8%	7.8%	1.4%	37%

화이트 베이스와 다르게 옐로우 베이스는 이미 그 자체로 크레마 젤라또라 부를 수 있는 고형분 함량을 가진다. 하지만 유지방의 풍미를 더 부여하고 노른자 특유의 비릿한 맛을 잡아주기 위해 레몬 제스트를 넣어 크레마 젤라또를 만들 수 있다.

Unlike the white base, the yellow base already has solids that can be called Crema gelato on its own. However, you can make Crema gelato by adding lemon zest to give the milk fat more flavor and neutralize the peculiar taste unique to egg yolks.

제조 방법 Procedure

1. 비커에 옐로우 베이스를 계량한다.
2. 옐로우 베이스에 생크림을 넣고 핸드블렌더로 믹싱한다.
3. 2에 레몬 제스트를 넣는다.
4. 제조기에 넣고 냉각 교반한다.
5. 바트 또는 카라피나에 추출한 후 급속 냉동고에 넣는다.
6. 바로 판매할 경우 5분간 급속 냉동한 후 쇼케이스로 옮긴다. 바로 판매하지 않을 경우(여유분일 경우) 1시간 급속 냉동한 후 -18°C 이하의 냉동고에 보관한다.

1. Measure the yellow base in a container.
2. Add cream to the base, and mix with a hand blender.
3. Add lemon zest into 2.
4. Pour into the machine and cold-churn the mixture.
5. Extract in a stainless-steel container or a carapina, then put in a blast freezer.
6. If you are selling it immediately, transfer it to the showcase after blast-freezing for 5 minutes. If not (if it is a spare), blast-freeze for one hour, then store frozen at -18°C or lower.

Gelato POD & PAC

(Potere Dolcificante) (Poter Anti Congelante)

젤라또 POD(감미도) & PAC(빙점강하력)

젤라또에 사용되는 당류의 역할을 정확히 알면 입 안에서 느껴지는 단맛과 질감을 자유롭게 조절할 수 있다. 당류 파트에서 이야기한 감미도를 'POD', 빙점강하력을 'PAC'라고 한다. 어떤 종류의 당을 사용하는지, 당 첨가 비율을 몇 %로 결정하는지에 따라 다양한 결과값을 도출할 수 있다.

POD와 PAC를 계산하기 위해서는 47~51p 표의 수치를 참고한다.

If you know the role of the sugars used in gelato precisely, you can freely adjust the sweetness and texture you feel in your mouth. The degree of sweetness I mentioned in the sugars section is called 'POD,' and the anti-freezing power is called 'PAC.' You can get various results depending on what type of sugar and what percentage is used.

To calculate POD and PAC, refer values on pages 47~51.

젤라또 POD Gelato POD

(Potere Dolcificante/ SP: Sweetening Power)

화이트 베이스 White Base

재료 Ingredients		중량 (g) Weight	당 (%) Sugar	지방 (%) Fat	무지유 고형분 (%) Non-fat milk solids	기타 고형분 (%) Other solids	총 고형분 (%) Total solids	POD (SP)	PAC (AFP)
우유 Milk		6,788	-	(3.5) 237.6	(9) 610.9	-	848.5	54.3 **(1)**	
생크림 Cream (38% fat)		957	-	(38) 363.7	(5.6) 53.6	-	417.3	6.1 **(2)**	
탈지분유 Skim milk powder		405	-	-	(96) 388.9	-	388.9	33 **(3)**	
설탕 Sugar		1,190	(100) 1,190	-	-	-	1,190	1,190 **(4)**	
함수결정포도당 Dextrose		185	(92) 170				170	122.4 **(5)**	
물엿 Glucose syrup		425	(80) 340	-	-	-	340	176.8 **(6)**	
복합안정제 Stabilizer		50	-	-	-	(100) 50	50	-	
TOTAL	g	10,000	1,700	601.3	1,053.4	50	3,404.7	1,582.6 **(7)**	
	%		17%	6%	10.5%	0.5%	34%	15.8 **(8)**	

1. **우유 (유당 5%)**

   유당 = 우유 중량 6,788 × 유당 5% = 339.4

   POD = 우유 유당 339.4 × 유당 POD 0.16 = 54.3 **(1)**

2. **생크림 (유당 4%)**

   유당 = 생크림 중량 957 × 유당 4% = 38.3

   POD = 생크림 유당 38.3 × 유당 POD 0.16 = 6.1 **(2)**

3. **탈지분유 (유당 51%)**

   유당 = 탈지분유 중량 405 × 유당 51% = 206.6

   POD = 탈지분유 유당 206.6 × 유당 POD 0.16 = 33 **(3)**

4. **설탕**

   POD = 설탕 당량 1,190 × 설탕 POD 1 = 1,190 **(4)**

5. **함수결정포도당**

   POD = 포도당 당량 170 × 포도당 POD 0.72 = 122.4 **(5)**

6. **물엿**

   POD = 물엿 당량 340 × 물엿 POD 0.52 = 176.8 **(6)**

7. 모든 POD 값을 더한다.

   POD = 우유 54.3 + 생크림 6.1 + 탈지분유 33 + 설탕 1,190 + 포도당 122.4 + 물엿 176.8 = 1,582.6 **(7)**

8. POD 1,582.6 ÷10,000%= 15.8 **(8)**

   15.8의 의미는 위 레시피에 실제로 들어간 당은 17%이지만 입 안에서 느껴지는 감미도는 15.8로, 1.2% 덜 달게 느껴진다는 의미이다.

   POD 개념을 잘 이해하면 사용하는 당과 당 비율에 따라 표면적인 당이 아닌 숨어 있는 단맛을 조절할 수 있다.

1. **Milk (5% lactose)**

   Lactose = Weight of milk 6,788 × 5% Lactose = 339.4

   POD = Lactose of milk 339.4 × Lactose POD 0.16 = 54.3 **(1)**

2. **Cream (4% lactose)**

   Lactose = Weight of cream 957 × 4% Lactose = 38.3

   POD = Lactose of cream 38.3 × Lactose POD 0.16 = 6.1 **(2)**

3. **Skim milk powder (51% Lactose)**

   Lactose = Weight of skim milk powder 405 × 51% Lactose = 206.6

   POD = Lactose of skim milk powder 206.6 × Lactose POD 0.16 = 33 **(3)**

4. **Sugar**

   POD = Sugar weight of sugar 1,190 × Sugar POD 1 = 1,190 **(4)**

5. **Dextrose**

   POD = Sugar weight of dextrose 170 × Dextrose POD 0.72 = 122.4 **(5)**

6. **Glucose syrup**

   POD = Sugar weight of glucose syrup 340 × Glucose syrup POD 0.52 = 176.8 **(6)**

7. Add all the POD values.

   POD = Milk 54.3 + Cream 6.1 + Skim milk powder 33 + Sugar 1,190 + Dextrose 122.4 + Glucose syrup 176.8 = 1,582.6 **(7)**

8. POD 1,582.6 ÷ 10,000% = 15.8 **(8)**

   The meaning of 15.8 is that the actual sugar in the recipe is 17%, but the sweetness you feel in the mouth is 15.8, which means it feels 1.2% less sweet.

   When you understand the concept of POD well, you can adjust the hidden sweetness rather than the actual sugar according to the sugar and sugar ratio used.

	POD	PAC
젤라또 Gelato	12 - 22	23 - 30
소르베또 Sorbetto	20 - 30	28 - 36

젤라또 PAC Gelato PAC

(Poter Anti Congelante/ AFP: Anti Freezing Point)

화이트 베이스 White Base

재료 Ingredients		중량 (g) Weight	당 (%) Sugar	지방 (%) Fat	무지유 고형분 (%) Non-fat milk solids	기타 고형분 (%) Other solids	총 고형분 (%) Total solids	POD (SP)	PAC (AFP)
우유 Milk		6,788	-	(3.5) 237.6	(9) 610.9	-	848.5	54.3	339.4 **(1)**
생크림 Cream (38% fat)		957	-	(38) 363.7	(5.6) 53.6	-	417.3	6.1	38.3 **(2)**
탈지분유 Skim milk powder		405	-	-	(96) 388.9	-	388.9	33	206.6 **(3)**
설탕 Sugar		1,190	(100) 1,190	-	-	-	1,190	1,190	1,190 **(4)**
함수결정포도당 Dextrose		185	(92) 170	-	-	-	170	122.4	323 **(5)**
물엿 Glucose syrup		425	(80) 340	-	-	-	340	176.8	312.8 **(6)**
복합안정제 Stabilizer		50	-	-	-	(100) 50	50	-	-
TOTAL	g	10,000	1,700	601.3	1,053.4	50	3,404.7	1,582.6	2,410.1 **(7)**
	%		17%	6%	10.5%	0.5%	34%	15.8	24.1 **(8)**

1. **우유 (유당 5%)**

   유당 = 우유 중량 6,788 × 유당 5% = 339.4

   PAC = 우유 유당 339.4 × 유당 PAC 1 = 339.4 **(1)**

2. **생크림 (유당 4%)**

   유당 = 생크림 중량 957 × 유당 4% = 38.3

   PAC = 생크림 유당 38.3 × 유당 PAC 1 = 38.3 **(2)**

3. **탈지분유 (유당 51%)**

   유당 = 탈지분유 중량 405 × 유당 51% = 206.6

   PAC = 탈지분유 유당 206.6 × 유당 PAC 1 = 206.6 **(3)**

4. **설탕**

   PAC = 설탕 당량 1,190 × 설탕 PAC 1 = 1,190 **(4)**

5. **함수결정포도당**

   PAC = 포도당 당량 170 × 포도당 PAC 1.9 = 323 **(5)**

6. **물엿**

   PAC = 물엿 당량 340 × 물엿 PAC 0.92 = 312.8 **(6)**

7. 모든 PAC값을 더한다.

   PAC = 우유 339.4 + 생크림 38.3 + 탈지분유 206.6 +
   설탕 1,190 + 포도당 323 + 물엿 312.8 = 2,410.1 **(7)**

8. PAC 2,410.1 ÷10,000%= 24.1 **(8)**

   쇼케이스 내에서 젤라또의 질감 변화에 영향을 미치는 요소는 다양하게 있지만 일차원적으로 젤라또의 적정 온도를 조절할 수 있는 방법은 아래와 같다.

   도출된 PAC값에 ÷ 2를 하여 -를 붙인다. 그러면 내 레시피와 어울리는 기본 쇼케이스 온도가 된다.

   이 경우 24.1 ÷ 2 = 12

   즉, -12°C가 이 레시피와 어울리는 쇼케이스 기본 설정 온도이다.

● 이 쇼케이스 온도는 참고 온도이고, 실제 환경에서는 발생하는 여러 변수에 따라 젤라띠에레가 상황에 맞는 온도로 조절할 수 있어야 한다.

1. **Milk (5% lactose)**

   Lactose = Weight of milk 6,788 × 5% Lactose = 339.4

   PAC = Lactose of milk 339.4 × Lactose PAC 1 = 339.4 **(1)**

2. **Cream (4% lactose)**

   Lactose = Weight of cream 957 × 4% Lactose = 38.3

   POD = Lactose of cream 38.3 × Lactose PAC 1 = 38.3 **(2)**

3. **Skim milk powder (51% Lactose)**

   Lactose = Weight of skim milk powder 405 × 51% Lactose = 206.6

   POD = Lactose of skim milk powder 206.6 × Lactose PAC 1 = 206.6 **(3)**

4. **Sugar**

   POD = Sugar weight of sugar 1,190 × Sugar PAC 1 = 1,190 **(4)**

5. **Dextrose**

   POD = Sugar weight of dextrose 170 × Dextrose PAC 1.9 = 323 **(5)**

6. **Glucose syrup**

   POD = Sugar weight of glucose syrup 340 × Glucose syrup PAC 0.92 = 312.8 **(6)**

7. Add all the PAC values.

   PAC = Milk 339.4 + Cream 38.3 + Skim milk powder 206.6 + Sugar 1,190 + Dextrose 323 + Glucose syrup 312.8 = 2,410.1 **(7)**

8. POD 2,410.1 ÷ 10,000% = 24.1 **(8)**

   Various factors affect the texture change of gelato within the showcase, but the method to adjust the appropriate temperature of gelato one-dimensionally is as follows.

   Divide the derived PAC value by 2 and add - (negative). This will be the default showcase temperature that matches the recipe.

   In the above example, 24.1 ÷ 2 = 12

   Therefore, -12°C is the default showcase temperature that matches this recipe.

● This showcase's temperature is a reference, and the gelatiere should be able to adjust it to a suitable temperature for the situation according to numerous variables that can occur in the actual environment.

Part 5.

Gelato Recipes

젤라또 레시피

1 Omija & Yogurt Gelato

Basic

오미자 & 요거트 젤라또

단맛, 신맛, 쓴맛, 짠맛, 매운맛을 느낄 수 있다고 하여 이름이 붙여진 오미자에 새콤달콤한 요거트를 더해 산뜻한 맛으로 즐길 수 있다.

Omija, which has a sweet, sour, bitter, salty, and spicy taste, as the name implies (five-flavor berry), can be enjoyed with a refreshing taste with sweet and slightly sour yogurt.

베이스를 사용한 레시피
Recipe using the base

Ingredients		Quantity
화이트 베이스	White base	250g
플레인 요거트	Plain yogurt	530g
오미자청	Omija syrup	200g
설탕	Sugar	8g
이눌린	Inulin	10g
복합안정제	Stabilizer	2g
TOTAL		1,000g

1. 비커에 화이트 베이스, 플레인 요거트, 오미자청을 계량한다.
2. 믹싱볼에 설탕, 이눌린, 복합안정제를 계량한 후 스패출러로 혼합한다.
3. 1에 2를 넣고 핸드블렌더로 믹싱한다.
4. 제조기에 넣고 냉각 교반한다.

1. Measure white base, plain yogurt, and omija syrup in a container.
2. Mix sugar, inulin, and stabilizer using a spatula.
3. Add 2 into 1 and mix with a hand blender.
4. Pour into the machine and cold-churn the mixture.

싱글 레시피
Single recipe

Ingredients		Quantity
우유	Milk	170
생크림	Cream (38% fat)	24g
물엿	Glucose syrup	10g
탈지분유	Skim milk powder	5g
설탕	Sugar	36g
함수결정포도당	Dextrose	5g
이눌린	Inulin	10g
복합안정제	Stabilizer	5g
플레인 요거트	Plain yogurt	530g
오미자청	Omija syrup	200g
TOTAL		1,000g

1. 냄비에 우유, 생크림, 물엿을 넣고 40°C로 가열한다.
2. 믹싱볼에 탈지분유, 설탕, 함수결정포도당, 이눌린, 복합안정제를 계량한 후 스패출러로 혼합한다.
3. 2를 1에 넣고 65~85°C로 가열한다.
4. 4°C로 냉각한다.
5. 0~12시간 숙성한다.
6. 5에 플레인 요거트, 오미자청을 넣고 핸드블렌더로 믹싱한다.
7. 제조기에 넣고 냉각 교반한다.

1. Heat milk, cream, and glucose syrup in a pot to 40°C.
2. Measure skim milk powder, sugar, dextrose, inulin, and stabilizer and combine with a spatula.
3. Add 2 into 1 and heat to 65~85°C.
4. Cool to 4°C.
5. Let rest for 0~12 hours.
6. Add plain yogurt and omija syrup into 5 and mix using a hand blender.
7. Pour into the machine and cold-churn the mixture.

당 Sugar	지방 Fat	무지유 고형분 Non-fat milk solids	기타 고형분 Other solids	총 고형분 Total solids	수분 Water
20.1%	3.6%	6.3%	1.8%	31.9%	68.1%

2 Cremino Pistachio Gelato

Basic

크레미노 피스타치오 젤라또

빈투바 초콜릿과 시칠리아산 피스타치오가 만나 풍부한 지방감과 재료 본연의 깊은 풍미를 느낄 수 있다.

Bean-to-bar chocolate and Sicilian pistachios are combined to create a rich creaminess and deep flavor of the ingredients.

베이스를 사용한 레시피
Recipe using the base

Ingredients		Quantity
화이트 베이스	White base	805g
다크초콜릿	Dark chocolate	60g
피스타치오 페이스트	Pistachio paste	100g
함수결정포도당	Dextrose	35g
TOTAL		1,000g

1. 비커에 화이트 베이스를 계량한다.
2. 믹싱볼에 녹인 다크초콜릿과 피스타치오 페이스트를 혼합한다.
3. 1에 2와 함수결정포도당을 넣고 핸드블렌더로 믹싱한다.
4. 제조기에 넣고 냉각 교반한다.

1. Measure white base in a container.
2. Combine melted dark chocolate and pistachio paste in a mixing bowl.
3. Add dextrose to 1 and 2; mix with a hand blender.
4. Pour into the machine and cold-churn the mixture.

싱글 레시피
Single recipe

Ingredients		Quantity
우유	Milk	610g
물엿	Glucose syrup	20g
탈지분유	Skim milk powder	50g
설탕	Sugar	125g
함수결정포도당	Dextrose	30g
복합안정제	Stabilizer	5g
피스타치오 페이스트	Pistachio paste	100g
다크초콜릿	Dark chocolate	60g
TOTAL		1,000g

1. 냄비에 우유와 물엿을 40°C로 가열한다.
2. 믹싱볼에 탈지분유, 설탕, 함수결정포도당, 복합안정제를 계량한 후 스패출러로 혼합한다.
3. 2를 1에 넣고 65~85°C로 가열한다.
4. 3에 피스타치오 페이스트와 다크초콜릿을 넣고 혼합한다.
5. 4°C로 냉각한다.
6. 0~12시간 숙성한다.
7. 제조기에 넣고 냉각 교반한다.

● 피스타치오 페이스트를 헤이즐넛 페이스트로 바꾸면 잔두야 맛으로 완성된다.

1. Heat milk and glucose syrup in a pot to 40°C.
2. Measure skim milk powder, sugar, dextrose, and stabilizer and combine with a spatula.
3. Add 2 into 1 and heat to 65~85°C.
4. Combine pistachio paste and dark chocolate to 3.
5. Cool to 4°C.
6. Let rest for 0~12 hours.
7. Pour into the machine and cold-churn the mixture.

● Replace pistachio paste to hazelnut paste to make Gianduia flavor.

당 Sugar	지방 Fat	무지유 고형분 Non-fat milk solids	기타 고형분 Other solids	총 고형분 Total solids	수분 Water
18.7%	10.1%	10%	6.8%	45.5%	54.5%

3 Mugwort & Cream Cheese Gelato

Basic

쑥 & 크림치즈 젤라또

최근 음료 업계에서 레트로 붐을 일으키며 전통 식재료로서 각광받고 있는 쑥을 젤라또에 담아보았다. 쑥 단독으로 만들어도 좋겠지만 특유의 강한 맛과 향을 부드럽게 중화시켜주기 위해 크림치즈를 곁들였다.

Recently, mugwort has been in the limelight as a traditional food ingredient due to the retro-boom in the beverage industry; and I tried using it in gelato. Using mugwort alone would be nice, but the cream cheese was added to neutralize its unique, pungent taste and aroma gently.

베이스를 사용한 레시피
Recipe using the base

Ingredients		Quantity
화이트 베이스	White base	875g
크림치즈	Cream cheese	70g
설탕	Sugar	35g
쑥가루	Mugwort powder	20g
쑥가루(토핑)	Mugwort powder (topping)	적당량 QS
TOTAL		1,000g

1. 비커에 모든 재료를 계량한 후 핸드블렌더로 믹싱한다.
2. 제조기에 넣고 냉각 교반한다.

● 쑥가루를 토핑해 서빙한다.

1. Measure all the ingredients in a container and mix with a hand blender.
2. Pour into the machine and cold-churn the mixture.

● Sprinkle with mugwort powder to serve.

싱글 레시피
Single recipe

Ingredients		Quantity
우유	Milk	595g
생크림	Cream (38% fat)	80g
물엿	Glucose syrup	40g
탈지분유	Skim milk powder	40g
설탕	Sugar	120g
함수결정포도당	Dextrose	60g
복합안정제	Stabilizer	5g
쑥가루	Mugwort powder	20g
크림치즈	Cream cheese	70g
쑥가루(토핑)	Mugwort powder (topping)	적당량 QS
TOTAL		1,000g

1. 냄비에 우유, 생크림, 물엿을 넣고 40°C로 가열한나.
2. 믹싱볼에 탈지분유, 설탕, 함수결정포도당, 복합안정제, 쑥가루를 계량한 후 스패출러로 혼합한다.
3. 2를 1에 넣고 65~85°C로 가열한다.
4. 4°C로 냉각한다.
5. 크림치즈를 넣고 핸드블렌더로 믹싱한다.
6. 0~12시간 숙성한다.
7. 제조기에 넣고 냉각 교반한다.

● 쑥가루를 토핑해 서빙한다.

1. Heat milk, cream, and glucose syrup in a pot to 40°C.
2. Measure skim milk powder, sugar, dextrose, mugwort powder, and stabilizer in a mixing bowl to combine.
3. Mix 2 with 1 and heat to 65~85°C.
4. Cool to 4°C.
5. Add cream cheese and mix with a hand blender.
6. Let rest for 0~12 hours.
7. Pour into the machine and cold-churn the mixture.

● Sprinkle with mugwort powder to serve.

당 Sugar	지방 Fat	무지유 고형분 Non-fat milk solids	기타 고형분 Other solids	총 고형분 Total solids	수분 Water
18.8%	7.1%	9.6%	0.7%	36.2%	63.8%

4 1101060 Gelato Basic

1101060 젤라또

아열대 과일인 그라비올라 잎의 효능과 사용할 수 있는 제품에 관한 연구는 많았지만, 과육을 사용한 연구 자료는 없어 대학원 석사 과정 때 그라비올라 과육을 첨가한 젤라또를 논제로 연구하고 학위를 받았다. 석사 학번(1101060)을 맛 이름으로 정하며 힘들었지만 뿌듯했던 연구를 한 페이지의 유쾌한 추억으로 남긴다. '신의 열매'라고 불릴 만큼 영양학적으로 우수한 효능을 가진 그라비올라는 해외에서 음료나 아이스크림으로 어렵지 않게 만나볼 수 있지만 한국에서는 아직 생소한 재료이다. 우유를 더해 부드럽게 표현했으며, 자칫 느끼할 수 있는 그라비올라 특유의 맛은 라임으로 산뜻하게 산미를 더해주었다.

There have been many studies on the efficacy of Graviola leaves, a subtropical fruit, and products used, but there was no research data using the pulp. So, during the master's degree course, I researched gelato made with Graviola pulp and received the degree. Naming it after my master's degree ID number (1101060), the challenging but proud research is left as a page of pleasant memory. Graviola has excellent nutritional effects and is called 'the fruit of the gods,' it's not difficult to find as a drink or ice cream overseas, but it is still an unfamiliar ingredient in Korea. I added softness by adding milk and refreshing acidity with lime to balance the unique taste of graviola, which some may feel greasy.

싱글 레시피
Single recipe

Ingredients		Quantity
우유	Milk	465g
물엿	Glucose syrup	30g
생크림	Cream (38% fat)	100g
탈지분유	Skim milk powder	50g
설탕	Sugar	110g
함수결정포도당	Dextrose	25g
라임즙	Lime juice	20g
그라비올라 과육	Graviola pulp	200g
라임제스트	Lime zest	적당량 QS
TOTAL		1,000g

1. 냄비에 우유, 물엿, 생크림을 넣고 40°C로 가열한다.
2. 믹싱볼에 탈지분유, 설탕, 함수결정포도당을 계량한 후 스패출러로 혼합한다.
3. 2를 1에 넣고 65~85°C로 가열한다.
4. 4°C로 냉각한다.
5. 4에 라임즙, 그라비올라 과육, 라임제스트를 넣고 핸드블렌더로 믹싱한다.
6. 0-12시간 숙성한다.
7. 제조기에 넣고 냉각 교반한다.

1. Heat milk, glucose syrup, and cream in a pot to 40°C.
2. Measure skim milk powder, sugar, and dextrose in a mixing bowl to combine.
3. Mix 2 with 1 and heat to 65~85°C.
4. Cool to 4°C.
5. Add lime juice, graviola pulp, and lime zest and mix using a hand blender.
6. Let rest for 0~12 hours.
7. Pour into the machine and cold-churn the mixture.

당 Sugar	지방 Fat	무지유 고형분 Non-fat milk solids	기타 고형분 Other solids	총 고형분 Total solids	수분 Water
19%	5.5%	9.3%	1.6%	35.3%	64.7%

5 Sujeonggwa & Chocolate Gelato

Basic

수정과 & 초콜릿 젤라또

2022년 싱가포르에서 열린 '아시안 젤라또 컵' 대회에서 한국적인 재료와 초콜릿을 조합한 메뉴를 선보이라는 미션을 받고 만들었던 메뉴다. 계피, 생강, 대추를 푹 끓인 후 배와 잣을 띄워 차갑게 마시는 한국 전통 음료인 수정과를 다크초콜릿과 혼합하여 수정과 특유의 쌉사래한 맛과 초콜릿의 달콤함이 잘 어우러지게 하였다.

This was created after receiving a mission to present a menu that combines Korean ingredients and chocolate at the Asian Gelato Cup held in Singapore in 2022. Sujeonggwa is a traditional Korean beverage served cold by boiling cinnamon, ginger, and jujube, topping with pears and pine nuts. We mixed it with dark chocolate to harmonize the unique bitterness of sujeonggwa with the sweetness of chocolate.

싱글 레시피
Single recipe

Ingredients		Weight
우유	Milk	250g
물엿	Glucose syrup	15g
생크림	Cream (38% fat)	60g
탈지분유	Skim milk powder	60g
설탕	Sugar	40g
함수결정포도당	Dextrose	10g
이눌린	Inulin	10g
복합안정제	Stabilizer	5g
수정과	Sujeonggwa (Cinnamon Punch)	450g
다크초콜릿	Dark chocolate	100g
TOTAL		1,000g

1. 냄비에 우유, 물엿, 생크림을 넣고 40°C로 가열한다.
2. 믹싱볼에 탈지분유, 설탕, 함수결정포도당, 이눌린, 복합안정제를 계량한 후 스패출러로 혼합한다.
3. 2를 1에 넣고 65~85°C로 가열한다.
4. 다크초콜릿을 넣고 섞어준다.
5. 4°C로 냉각한다.
6. 수정과를 넣고 0~12시간 숙성한다.

- 수정과는 시판 제품 중 기호에 맞는 것을 사용한다. 여기에서는 콩그린식품 장모님 수정과 제품을 사용했다.

7. 제조기에 넣고 냉각 교반한다.

1. Combine milk, glucose syrup, and cream; heat to 40°C.
2. Measure skim milk powder, sugar, dextrose, inulin, and stabilizer in a mixing bowl to combine.
3. Mix 2 with 1 and heat to 65~85°C.
4. Mix with dark chocolate.
5. Cool to 4°C.
6. Mix with sujeonggwa and let rest for 0~12 hours.

- Use the commercially available sujeongwa that suits your taste. I used KongGreen Food's Mother-in-law Sujeonggwa.

7. Pour into the machine and cold-churn the mixture.

당 Sugar	지방 Fat	무지유 고형분 Non-fat milk solids	기타 고형분 Other solids	총 고형분 Total solids	수분 Water
19.9%	5.4%	8.2%	2%	35.4%	64.6%

6 Pinoli & Riso Gelato

Topping

피놀리 & 리조 젤라또

보통 젤라또에서 사용하는 리조(쌀)는 쌀알의 식감이 잘 느껴지도록 만들고, 우유 맛의 기본 젤라또에 많이 사용된다. 하지만 우유 맛이 아닌 젤라또에 사용해도 리조가 가진 맛과 식감을 충분히 살릴 수 있다. 여기에서는 개인적으로 참 좋아하는 조합인 고소한 황잣 젤라또에 리조를 더해보았다.

Riso (rice), which is usually used in gelato, should be made to feel the texture of the rice grains and is often used in basic milk-flavored gelato. However, even if it's not used for non-milk gelato, you can fully enjoy the taste and texture of riso. Here, I added riso to my favorite rich and nutty yellow pine nut gelato.

가당 쌀
Sweetened rice

Ingredients		Quantity
불린 쌀	Soaked rice	40g
물	Water	400g
설탕	Sugar	50g
함수결정포도당	Dextrose	10g
TOTAL		500g

1. 쌀은 물에 충분히 불려 준비한다.
2. 모든 재료를 냄비에 넣고 중약불에서 가열한다.
3. 스패출러로 저어가며 죽 같은 상태가 될 때까지 졸인다.

- 원하는 향신료를 넣고 같이 졸이면 쌀에 풍미가 더해져 또 다른 맛의 가당 쌀을 만들 수 있다.
- 가당 쌀에서 유지방 풍미가 느껴지기를 원한다면 물 대신 우유를 사용한다.

1. Soak the rice sufficiently in water.
2. Put all the ingredients in a pot and cook over medium-low heat.
3. Simmer until it resembles porridge while stirring continuously.

- Cook with desired spices to make different flavors of sweetened rice by adding flavor to the rice.
- If you want the sweetened rice to have a dairy flavor, use milk instead of water.

베이스를 사용한 레시피

Recipe using the base

Ingredients		Quantity
화이트 베이스	White base	855g
생크림	Cream (38% fat)	45g
꿀	Honey	50g
구운 황잣	Roasted yellow pine nuts	50g
가당 쌀	Sweetened rice	적당량 QS
TOTAL		1,000g

1. 비커에 화이트 베이스, 생크림, 꿀, 구운 황잣을 넣고 핸드블렌더로 믹싱한다.
2. 제조기에 넣고 냉각 교반한다.
3. 바트에 추출하며 가당 쌀을 섞어준다.

1. Put white base, cream, honey, and roasted yellow pine nuts in a container and mix using a hand blender.
2. Pour into the machine and cold-churn the mixture.
3. Extract into a stainless-steel container and mix with sweetened rice.

싱글 레시피
Single recipe

Ingredients		Quantity
우유	Milk	680g
생크림	Cream (38% fat)	40g
꿀	Honey	40g
물엿	Glucose syrup	35g
탈지분유	Skim milk powder	30g
설탕	Sugar	120g
복합안정제	Stabilizer	5g
구운 황잣	Roasted yellow pine nuts	50g
가당 쌀	Sweetened rice	적당량 QS
TOTAL		1,000g

1. 냄비에 우유, 생크림, 꿀, 물엿을 넣고 40°C로 가열한다.
2. 믹싱볼에 탈지분유, 설탕, 복합안정제를 계량한 후 스패출러로 혼합한다.
3. 2를 1에 넣고 65~85°C로 가열한다.
4. 4°C로 냉각한다.
5. 구운 황잣을 넣고 핸드블렌더로 믹싱한다.
6. 0~12시간 숙성한다.
7. 제조기에 넣고 냉각 교반한다.
8. 바트에 추출하며 가당 쌀을 섞어준다.

1. Heat milk, cream, honey, and glucose syrup in a pot to 40°C.
2. Measure skim milk powder, sugar, and stabilizer in a mixing bowl to combine.
3. Mix 2 with 1 and heat to 65 85°C.
4. Cool to 4°C.
5. Add roasted yellow pine nuts and mix with a hand blender.
6. Let rest for 0~12 hours.
7. Pour into the machine and cold-churn the mixture.
8. Extract into a stainless-steel container and mix with sweetened rice.

당 Sugar	지방 Fat	무지유 고형분 Non-fat milk solids	기타 고형분 Other solids	총 고형분 Total solids	수분 Water
18.7%	7.3%	9.6%	0.5%	36.1%	63.9%

Memo.

7 # Zuppa Inglese Gelato

Topping

주빠 잉글레제 젤라또

주빠 잉글레제는 'Zuppa(스프)'와 'Inglese(영국)'의 합성어, 즉 '영국의 스프'라는 뜻으로 영국 디저트인 'Trifle(트라이플)'에서 영감을 받아 만들어진 메뉴다. 주빠 잉글레제는 알케르메스 술로 만든 시럽을 스펀지 시트나 사보이아르디 과자에 적셔 크림과 초콜릿을 층층이 발라 떠먹는 세미프레도(semifreddo, 고체와 액체 중간 형태를 띄는 이탈리아 아이스크림)이지만 여기에서는 젤라또로 표현해보았다.

Zuppa Ingelse is a combined word of 'Zuppa (soup)' and 'Ingelese (England),' which means 'British soup' and is a menu inspired by the British dessert 'Trifle.' Zuppa Inglese is a semifreddo (Italian ice cream with a texture between solid and liquid) that is eaten by soaking a sponge cake or Savoiardi cookie with syrup made from Alchermes liqueur and laying it with cream and chocolate, but here I created it into gelato.

옐로우 소스*
Yellow sauce

Ingredients		Quantity
물	Water	130g
우유	Milk	380g
생크림	Cream (38% fat)	50g
설탕	Sugar	235g
노른자	Egg yolks	205g
TOTAL		1,000g

1. 비커에 모든 재료를 넣고 휘퍼로 혼합한다.
2. 65°C로 가열한 후 냉장 보관하여 사용한다.

● 옐로우 베이스를 따로 만들지 않고 화이트 베이스에 옐로우 소스를 섞어주면 옐로우 베이스로 만들 수 있는 다양한 젤라또 맛을 만들 수 있다.

● **보관 기간**
옐로우 베이스: 냉장 1주일
옐로우 소스: 냉장 15일

1. Add all the ingredients and combine with a whisk.
2. Heat to 65°C and refrigerate to use.

● You can mix the yellow sauce to the white base instead of making a separate yellow base with make various flavors that can be made with the yellow base.

● **STORAGE PERIOD**
Yellow base: 1 week, refrigerated
Yellow sauce: 15 days, refrigerated

알케르메스 시럽*
Alchermes syrup

Ingredients		Quantity
물	Water	550g
설탕	Sugar	280g
레몬제스트	Zest of one lemon	레몬 1개 분량
알케르메스	Alchermes	170g
TOTAL		1,000g

1. 냄비에 물, 설탕, 레몬제스트를 넣고 설탕이 녹을 때까지 끓인 후 식힌다.
2. 1이 식으면 알케르메스를 넣고 섞는다.

1. Bring water, sugar, and lemon zest to a boil until the sugar dissolves and let cool.
2. Once 1 is cooled, mix with Alchermes.

베이스를 사용한 레시피
Recipe using the base

Ingredients		Quantity
화이트 베이스	White base	700g
옐로우 소스*	Yellow sauce*	300g
초콜릿 시럽	Chocolate syrup	적당량 QS
알케르메스 시럽*에 적신 사보이아르디	Savoiardi cookie soaked with Alchermes syrup*	적당량 QS
TOTAL		1,000g

1. 비커에 화이트 베이스, 옐로우 소스를 계량한다.
2. 1을 핸드블렌더로 믹싱한 후 제조기에 넣고 냉각 교반한다.
3. 바트에 추출하며 초콜릿 시럽과 알케르메스 시럽에 적신 사보이아르디를 넣는다.

1. Measure white base and yellow sauce in a container.
2. Mix 1 using a hand blender and pour into the machine and cold-churn the mixture.
3. Extract in a stainless-steel container and add chocolate syrup and Savoiardi cookies soaked with Alchermes syrup.

싱글 레시피
Single recipe

Ingredients		Quantity
우유	Milk	635g
물엿	Glucose syrup	20g
생크림	Cream (38% fat)	70g
노른자	Egg yolks	65g
탈지분유	Skim milk powder	30g
설탕	Sugar	145g
함수결정포도당	Dextrose	30g
복합안정제	Stabilizer	5g
초콜릿 시럽	Chocolate syrup	적당량 QS
알케르메스 시럽*에 적신 사보이아르디	Savoiardi cookies soaked with Alchermes syrup*	적당량 QS
TOTAL		1,000g

1. 냄비에 우유, 물엿, 생크림, 노른자를 넣고 40°C로 가열한다.
2. 믹싱볼에 탈지분유, 설탕, 함수결정포도당, 복합안정제를 계량한 후 스패출러로 혼합한다.
3. 2를 1에 넣고 65°C로 가열한다.
4. 4°C로 냉각한다.
5. 0~12시간 숙성한다.
6. 제조기에 넣고 냉각 교반한다.
7. 바트에 추출하며 초콜릿 시럽, 알케르메스 시럽에 적신 사보이아르디를 넣는다.

1. Heat milk, glucose syrup, cream, and egg yolks in a pot to 40°C.
2. Measure skim milk powder, sugar, dextrose, and stabilizer in a mixing bowl to combine.
3. Add 2 into 1 and heat to 65°C.
4. Cool to 4°C.
5. Let rest for 0~12 hours.
6. Pour into the machine and cold-churn the mixture.
7. Extract in a stainless-steel container and add chocolate syrup and Savoiardi cookies soaked with Alchermes syrup.

당 Sugar	지방 Fat	무지유 고형분 Non-fat milk solids	기타 고형분 Other solids	총 고형분 Total solids	수분 Water
18.9%	6.9%	8.6%	1.7%	36%	64%

Memo.

8 Tiramisu Gelato Topping

티라미수 젤라또

티라미수는 이탈리아어로 '나를 끌어올려줘'라는 뜻의 디저트로 에스프레소, 카카오파우더, 마스카르포네, 사보이아르디로 만드는 이탈리아의 전통 세미프레도(semifreddo, 고체와 액체 중간 형태를 띄는 이탈리아 아이스크림)이다. 여기에서는 세미프레도를 만드는 방법과 동일하게 젤라또로 표현해보았다.

Tiramisu is a dessert that means 'lift me up' in Italian and is a traditional Italian semifreddo (Italian ice cream with a texture between solid and liquid) made with espresso, cacao powder, mascarpone, and Savoiardi. I made gelato in the same method as making semifreddo.

커피 시럽*
Coffee syrup

Ingredients		Quantity
물	Water	600g
설탕	Sugar	350g
인스턴트 커피가루	Instant coffee powder	50g
TOTAL		1,000g

1. 냄비에 물, 설탕, 인스턴트 커피가루를 넣고 녹을 때까지 끓인다.
2. 충분히 식힌 후 사용한다.

1. Boil water, sugar, and instant coffee powder in a saucepan.
2. Cool sufficiently before use.

옐로우 소스*
Yellow sauce

Ingredients		Quantity
물	Water	130g
우유	Milk	380g
생크림	Cream (38% fat)	50g
설탕	Sugar	235g
노른자	Egg yolks	205g
TOTAL		1,000g

1. 비커에 모든 재료를 넣고 휘퍼로 혼합한다.
2. 65°C로 가열한 후 냉장 보관하여 사용한다.

- 옐로우 베이스를 따로 만들지 않고 화이트 베이스에 옐로우 소스를 섞어주면 옐로우 베이스로 만들 수 있는 다양한 젤라또 맛을 만들 수 있다.

- **보관 기간**
  옐로우 베이스: 냉장 1주일
  옐로우 소스: 냉장 15일

1. Add all the ingredients and combine with a whisk.
2. Heat to 65°C and refrigerate to use.

- You can mix the yellow sauce to the white base instead of making a separate yellow base with make various flavors that can be made with the yellow base.

- **STORAGE PERIOD**
  Yellow base: 1 week, refrigerated
  Yellow sauce: 15 days, refrigerated

베이스를 사용한 레시피

Recipe using the base

Ingredients		Quantity
화이트 베이스	White base	690g
옐로우 소스*	Yellow sauce*	210g
마스카르포네	Mascarpone	80g
설탕	Sugar	17g
인스턴트 커피가루	Instant coffee powder	3g
카카오파우더	Cocoa powder	적당량 QS
커피 시럽*에 적신 사보이아르디	Savoiardi cookies soaked with coffee syrup*	적당량 QS
TOTAL		1,000g

1. 비커에 화이트 베이스, 옐로우 소스, 마스카르포네, 설탕, 인스턴트 커피가루를 계량한 후 핸드블렌더로 믹싱한다.
2. 제조기에 넣고 냉각 교반한다.
3. 바트에 추출하며 카카오파우더와 커피 시럽에 적신 사보이아르디를 넣는다.

1. Measure white base, yellow sauce, mascarpone, sugar, and instant coffee powder in a container and mix with a hand blender.
2. Pour into the machine and cold-churn the mixture.
3. Extract in a stainless-steel container and add cocoa powder and Savoiardi cookies soaked with coffee syrup.

커피 시럽에 적신 사보이아르디

Savoiardi cookies soaked with coffee syrup

싱글 레시피
Single recipe

Ingredients		Quantity
우유	Milk	577g
물엿	Glucose syrup	20g
생크림	Cream (38% fat)	80g
노른자	Egg yolks	40g
탈지분유	Skim milk powder	30g
설탕	Sugar	140g
함수결정포도당	Dextrose	30g
복합안정제	Stabilizer	5g
커피가루	Coffee powder	40g
마스카르포네	Mascarpone	75g
카카오파우더	Cocoa powder	적당량 QS
커피 시럽*에 적신 사보이아르디	Savoiardi cookies soaked with coffee syrup*	적당량 QS
TOTAL		1,000g

1. 냄비에 우유, 물엿, 생크림, 노른자를 넣고 40°C로 가열한다.
2. 믹싱볼에 탈지분유, 설탕, 함수결정포도당, 복합안정제, 커피가루를 계량한 후 스패출러로 혼합한다.
3. 2를 1에 넣고 65°C로 가열한다.
4. 4°C로 냉각한다.
5. 0~12시간 숙성한다.
6. 5에 마스카르포네를 넣고 핸드블렌더로 믹싱한다.
7. 제조기에 넣고 냉각 교반한다.
8. 바트에 추출하며 카카오파우더와 커피 시럽에 적신 사보이아르디를 넣는다.

1. Heat milk, glucose syrup, cream, and egg yolks in a pot to 40°C.
2. Measure skim milk powder, sugar, dextrose, stabilizer, and coffee powder in a mixing bowl to combine.
3. Add 2 into 1 and heat to 65°C.
4. Cool to 4°C.
5. Let rest for 0~12 hours.
6. Add mascarpone to 5 and mix using a hand blender.
7. Pour into the machine and cold-churn the mixture.
8. Extract in a stainless-steel container and add cocoa powder and Savoiardi cookies soaked with coffee syrup.

당 Sugar	지방 Fat	무지유 고형분 Non-fat milk solids	기타 고형분 Other solids	총 고형분 Total solids	수분 Water
18.7%	9.4%	8.5%	1.2%	37.8%	62.2%

9 Panettone Gelato

Topping

파네토네 젤라또

이탈리아 전통 디저트 중 하나인 '파네토네Panettone'는 주로 크리스마스 시즌에 즐겨 먹는 디저트이다. 젤라또에서 파네토네의 풍미를 극대화시키기 위해 파네토네를 직접 인퓨징하였고, 건포도와 오렌지필을 넣어 파네토네를 먹었을 때 느껴지는 당절임 과일의 맛과 쫀득한 식감을 표현했다.

'Panettone' is one of the traditional Italian desserts that is mainly consumed during the Christmas season. To maximize the flavor of panettone in gelato, I infused the panettone itself, and raisins and orange peels were added to optimize the taste of candied fruits and the texture you feel when eating panettone.

베이스를 사용한 레시피
Recipe using the base

Ingredients		Quantity
화이트 베이스	White base	750g
우유	Milk	100g
파네토네	Panettone	150g
건포도	Raisins	적당량 QS
오렌지필	Orange peel	적당량 QS
TOTAL		1,000g

1. 비커에 화이트 베이스, 우유, 파네토네를 계량한 후 냉장고에 12~24시간 두어 파네토네의 향을 우려낸다.
2. 1의 파네토네 2/3은 건져내고, 나머지는 핸드블렌더로 믹싱한다.
3. 거름망을 사용해 걸러낸다.
4. 제조기에 넣고 냉각 교반한다.
5. 바트에 추출하며 당절임 건포도와 오렌지필을 넣고 혼합한다.

1. Measure white base, milk, and panettone in a beaker. Refrigerate for 12~24 hours to extract the flavor.
2. Take out 1/2 of the panettone, and mix the rest with a hand blender.
3. Filter through a sieve.
4. Pour into the machine and cold-churn the mixture.
5. Extract into a stainless-steel container and mix with sweetened raisins and orange peel.

싱글 레시피
Single recipe

Ingredients		Quantity
우유	Milk	585g
생크림	Cream (38% fat)	70g
물엿	Glucose syrup	20g
탈지분유	Skim milk powder	40g
설탕	Sugar	100g
함수결정포도당	Dextrose	30g
복합안정제	Stabilizer	5g
파네토네	Panettone	150g
건포도	Raisins	적당량 QS
오렌지필	Orange peel	적당량 QS
TOTAL		1,000g

1. 냄비에 우유, 생크림, 물엿을 넣고 40°C로 가열한다.
2. 믹싱볼에 탈지분유, 설탕, 함수결정포도당, 복합안정제를 계량한 후 스패출러로 혼합한다.
3. 2를 1에 넣고 65~85°C로 가열한다.
4. 4°C로 냉각한다.
5. 0~12시간 숙성한다.
6. 파네토네를 5에 넣고 12~24시간 동안 향을 우려낸다.
7. 파네토네 2/3는 건져내고, 나머지는 핸드블렌더로 믹싱한다.
8. 거름망을 사용해 걸러준다.
9. 제조기에 넣고 냉각 교반한다.
10. 바트에 추출하며 당절임 건포도와 오렌지필을 넣고 혼합한다.

1. Heat milk, cream, glucose syrup in a pot to 40°C.
2. Measure skim milk powder, sugar, dextrose, and stabilizer in a mixing bowl and combine with a spatula.
3. Mix 2 with 1 and heat to 65~85°C.
4. Cool to 4°C.
5. Let rest for 0~12 hours.
6. Add panettone to 5 and let it infuse for 12~24 hours.
7. Take out 2/3 of the panettone, and combine the rest using a hand blender.
8. Filter through a sieve.
9. Pour into the machine and cold-churn the mixture.
10. Extract into a stainless-steel container and mix with sweetened raisins and orange peel.

당 Sugar	지방 Fat	무지유 고형분 Non-fat milk solids	기타 고형분 Other solids	총 고형분 Total solids	수분 Water
18%	6.4%	9.2%	2.5%	36.1%	63.9%

10 Apple Crumble Gelato

Topping

애플 크럼블 젤라또

크럼블이 가진 버터의 향을 차가운 온도에서도 느낄 수 있도록 브라운 버터를 만들어 젤라또에 더했다. 계피가루를 넣어 만든 달콤한 크럼블과 아삭하게 씹히는 식감이 매력적인 사과 포셰를 토핑으로 곁들였다.

I made brown butter and added it to the gelato so that the crumble's butter aroma exists even at a cold temperature. Sweet crumbles made with cinnamon powder and apple pochée with a crispy bite are topped for a charming texture.

크럼블
Crumbles

Ingredients		Quantity
박력분	Cake flour	110g
아몬드가루	Almond powder	45g
버터	Butter	80g
설탕	Sugar	60g
소금	Salt	3g
계피가루	Ground cinnamon	2g
TOTAL		300g

1. 푸드프로세서에 모든 재료를 계량한 후 갈아준다.
2. 사각팬에 덩어리지도록 흩뿌린다.
3. 175°C로 예열된 오븐에서 25분간 구워 준다.

● 오븐의 사양에 따라 시간과 온도를 조절한다.

1. Measure all the ingredients and combine in a food processor.
2. Sprinkle on a baking tray in lumps.
3. Bake for 25 minutes in an oven preheated to 175°C.

● Adjust the baking time and temperature according to the oven's specifications.

베이스를 사용한 레시피
Recipe using the base

Ingredients		Quantity
브라운 버터	Brown butter	35g
우유	Milk	45g
옐로우 베이스	Yellow base	910g
설탕	Sugar	10g
레몬제스트	Zest of one lemon	레몬 1개 분량
사과 포셰	Apple pochée	적당량 QS
크럼블	Crumbles	적당량 QS
TOTAL		1,000g

• 사과 포셰는 97p를 참고한다.
• Refer to p.97 for apple pochée.

1. 냄비에 버터를 넣고 가열해 진한 브라운 색으로 태워 브라운 버터를 만든다.
2. 완성된 브라운 버터를 고운 체망에 걸러 35g을 계량한 후 따뜻하게 데운 우유와 섞어준다.

• 버터 60g을 태우면 약 35~40g의 브라운 버터가 만들어진다.

3. 비커에 2와 옐로우 베이스, 설탕, 레몬제스트를 넣고 섞는다.
4. 제조기에 넣고 냉각 교반한다.
5. 바트에 추출하며 사과 포셰와 크럼블을 섞어준다.

• 사과의 쫀득한 식감을 원하는 경우 포셰 방식이 아닌 냄비에 졸이는 방식으로 만든다.

1. Heat butter in a pot and cook until it turns dark brown to make brown butter.
2. Strain the brown butter through a fine sieve and measure 35 grams. Mix with warmed milk.

• Cooking 60 grams of butter makes about 35~40 grams of brown butter.

3. Combine 2, yellow base, sugar, and lemon zest in a container.
4. Pour into the machine and cold-churn the mixture.
5. Extract into a stainless-steel container and mix with apple pochée and crumbles.

• If you want chewy textured apples, simmer them in a pot rather than using the pochée method.

브라운 버터 brown butter

싱글 레시피
Single recipe

Ingredients		Quantity
브라운 버터	Brown butter	40g
우유	Milk	600g
물엿	Glucose syrup	30g
생크림	Cream (38% fat)	70g
노른자	Egg yolks	60g
레몬제스트	Zest of one lemon	레몬 1개 분량
탈지분유	Skim milk powder	30g
설탕	Sugar	125g
함수결정포도당	Dextrose	40g
복합안정제	Stabilizer	5g
사과 포셰	Apple pochée	적당량 QS
크럼블	Crumbles	적당량 QS
TOTAL		1,000g

- 사과 포셰는 97p를 참고한다.
- Refer to p.97 for apple pochée.

1. 냄비에 버터를 넣고 가열해 진한 브라운 색으로 태워 브라운 버터를 만든다.
2. 완성된 브라운 버터를 고운 체망에 걸러 40g을 계량한다.
3. 냄비에 우유, 물엿, 생크림, 노른자, 브라운 버터, 레몬제스트를 넣고 40°C로 가열한다.
4. 믹싱볼에 탈지분유, 설탕, 함수결정포도당, 복합안정제를 계량한 후 스패출러로 혼합한다.
5. 4를 3에 넣고 65~85°C로 가열한다.
6. 4°C로 냉각한다.
7. 0~12시간 숙성한다.
8. 제조기에 넣고 냉각 교반한다.
9. 바트에 추출하며 사과 포셰와 크럼블을 섞어준다.

1. Heat butter in a pot and cook until it turns dark brown to make brown butter.
2. Strain the brown butter through a fine sieve and measure 40 grams.
3. Heat milk, glucose syrup, cream, egg yolks, brown butter, and lemon zest to 40°C.
4. Measure skim milk powder, sugar, dextrose, and stabilizer in a mixing bowl to combine.
5. Mix **4** with **3** and heat to 65~85°C
6. Cool to 4°C.
7. Let rest for 0~12 hours.
8. Pour into the machine and cold-churn the mixture.
9. Extract into a stainless-steel container and mix with apple pochée and crumbles.

당 Sugar	지방 Fat	무지유 고형분 Non-fat milk solids	기타 고형분 Other solids	총 고형분 Total solids	수분 Water
18.6%	9.9%	8.3%	1.6%	38.4%	61.6%

11 Ricotta & Fig Gelato

Topping

리코타 & 무화과 젤라또

리코타 특유의 산미와 캐러멜라이징한 무화과의 달콤함이 조화로운 젤라또. 부드러운 질감 속에서 느껴지는 무화과의 식감 또한 매력적이다.

It is a gelato with the unique acidity of ricotta and the sweetness of caramelized figs. The texture of figs in the soft gelato is one of a kind.

캐러멜라이즈한 무화과*
Caramelized figs

Ingredients		Quantity
무화과	Figs	900g
설탕	Sugar	100g
TOTAL		1,000g

1. 오븐팬에 무화과를 펼치고 설탕을 골고루 뿌린다.
2. 170°C로 예열된 오븐에서 약 30분간 굽는다.

● 오븐의 사양에 따라 시간 및 온도 조절이 필요하다.

1. Evenly spread the figs on a baking tray and sprinkle with sugar.
2. Cook for 30 minutes in an oven preheated to 170°C.

● Cooking time and temperature should be adjusted according to the specification of the oven.

베이스를 사용한 레시피
Recipe using the base

Ingredients		Quantity
화이트 베이스	White base	780g
리코타	Ricotta	160g
설탕	Sugar	60g
캐러멜라이즈한 무화과*	Caramelized figs*	적당량 QS
TOTAL		1,000g

1. 비커에 화이트 베이스, 리코타, 설탕을 계량한다.
2. 1을 핸드블렌더로 믹싱한다.
3. 제조기에 넣고 냉각 교반한다.
4. 바트에 추출하며 캐러멜라이즈한 무화과를 넣는다.

1. Measure white base, ricotta, and sugar in a container.
2. Mix using a hand blender.
3. Pour into the machine and cold-churn the mixture.
4. Extract in a stainless-steel container and add caramelized figs.

싱글 레시피
Single recipe

Ingredients		Quantity
우유	Milk	520g
물엿	Glucose syrup	30g
생크림	Cream (38% fat)	75g
탈지분유	Skim milk powder	40g
설탕	Sugar	150g
함수결정포도당	Dextrose	20g
복합안정제	Stabilizer	5g
리코타	Ricotta	160g
캐러멜라이즈한 무화과*	Caramelized figs*	적당량 QS
TOTAL		1,000g

1. 냄비에 우유, 물엿, 생크림을 넣고 40°C로 가열한다.
2. 믹싱볼에 탈지분유, 설탕, 함수결정포도당, 복합안정제를 계량한 후 스패출러로 혼합한다.
3. 2를 1에 넣고 65~85°C로 가열한다.
4. 4°C로 냉각한다.
5. 0~12시간 숙성한다.
6. 5에 리코타를 넣고 핸드블렌더로 믹싱한다.
7. 제조기에 넣고 냉각 교반한다.
8. 바트에 추출하며 캐러멜라이즈한 무화과를 섞는다.

1. Heat milk, glucose syrup, and cream in a pot to 40°C.
2. Measure skim milk powder, sugar, dextrose, and stabilizer in a mixing bowl to combine.
3. Add **2** into **1** and heat to 65°C.
4. Cool to 4°C.
5. Let rest for 0~12 hours.
6. Add ricotta to **5** and mix using a hand blender.
7. Pour into the machine and cold-churn the mixture.
8. Extract in a stainless-steel container and add caramelized figs.

당 Sugar	지방 Fat	무지유 고형분 Non-fat milk solids	기타 고형분 Other solids	총 고형분 Total solids	수분 Water
19.3%	6.8%	10.4%	0.5%	37%	63%

Memo.

12

Hummingbird Gelato

Topping

허밍버드 젤라또

미국의 전통 케이크인 허밍버드는 바나나, 파인애플, 피칸이 들어가 과일의 달콤함과 견과류의 고소함이 어우러진 디저트이다. 일뿌 젤라또 수업에서는 수강생이 원하는 맛의 조합으로 실습을 진행하고 있는데, 해외에서 오신 수강생 분이 허밍버드 케이크 맛을 젤라또로 표현하고 싶다고 하셔서 만들게 된 메뉴다.

Hummingbird cake is an American dessert that combines the sweetness of bananas, pineapples, and nutty pecan. In il più Gelato class, students practice with the combination of flavors they want, and one of the students from abroad wanted to recreate the taste of Hummingbird cake, so we did.

싱글 레시피
Single recipe

Ingredients		Quantity
브라운 버터	Brown butter	30g
우유	Milk	455g
생크림	Cream (38% fat)	30g
물엿	Glucose syrup	40g
탈지분유	Skim milk powder	55g
설탕	Sugar	120g
함수결정포도당	Dextrose	20g
바나나	Banana	250g
캔디드 피칸	Candied pecans	적당량 QS
파인애플 포셰	Pineapple pochée	적당량 QS
TOTAL		1,000g

• 파인애플 포셰는 97p를 참고한다.

• Refer to p.97 for Pineapple pochée.

브라운 버터 brown butter

캔디드 피칸 Candied pecans

1. 냄비에 버터를 넣고 가열해 진한 브라운 색으로 태워 브라운 버터를 만든다.
2. 완성된 브라운 버터를 고운 체망에 걸러 30g을 계량한다.
3. 냄비에 우유, 생크림, 물엿, 브라운 버터를 넣고 40°C로 가열한다.
4. 믹싱볼에 탈지분유, 설탕, 함수결정포도당을 계량한 후 스패출러로 혼합한다.
5. 4를 3에 넣고 65~85°C로 가열한다.
6. 4°C로 냉각한다.
7. 0~12시간 숙성한다.
8. 7에 바나나를 넣고 핸드블렌더로 믹싱한다.
9. 제조기에 넣고 냉각 교반한다.
10. 바트에 추출하며 캔디드 피칸과 파인애플 포셰를 섞어준다.

• 캔디드 피칸 제조는 '고르곤졸라 & 피스타치오 젤라또(185p)'의 캔디드 피스타치오와 동일하다.

1. Heat butter in a pot and cook until it turns dark brown to make brown butter.
2. Strain the brown butter through a fine sieve and measure 30 grams.
3. Heat milk, cream, glucose syrup, and brown butter in a pot to 40°C.
4. Measure skim milk powder, sugar, and dextrose in a mixing bowl to combine.
5. Mix 4 with 3 and heat to 65~85°C.
6. Cool to 4°C.
7. Let rest for 0~12 hours.
8. Add banana to 7 and mix with a hand blender.
9. Pour into the machine and cold-churn the mixture.
10. Extract into a stainless-steel container and mix with candied pecans and pineapple pochée.

• Making candied pecans is the same as candied pistachio in 'Gorgonzola & Pistachio Gelato (p.185)'.

당 Sugar	지방 Fat	무지유 고형분 Non-fat milk solids	기타 고형분 Other solids	총 고형분 Total solids	수분 Water
20.5%	5.3%	9.3%	2.3%	37.4%	62.6%

13 Almond & Mint Gelato

Hot Infusion

아몬드 & 민트 젤라또

2022년 싱가포르에서 열린 '아시안 젤라또 컵' 대회에서 한국 대표 선수들이 선보인 맛. 아몬드를 우유에 갈아 고소함을 배가시켰다. 여기에 향긋한 페퍼민트, 싱그러운 초록빛을 더해주는 민트 페이스트, 상큼한 라임을 매칭해 대회의 주제였던 '트로피컬 파라다이스'의 맛을 표현했다.

It was presented by the Korean National team at the Asian Gelato Cup held in Singapore in 2022. Almonds are ground with milk to amplify the nuttiness. Here, fragrant peppermint, mint paste to add a lush green color, and refreshing lime are matched to express the contest's theme flavor, 'Tropical Paradise.'

베이스를 사용한 레시피
Recipe using the base

Ingredients		Quantity
구운 아몬드	Roasted almonds	250g
화이트 베이스	White base	890g
페퍼민트 잎	Peppermint leaves	적당량 QS
생크림	Cream (38% fat)	45g
설탕	Sugar	30g
민트 페이스트	Mint paste	5g
라임즙	Lime juice	30g
TOTAL		1,000g

1. 아몬드는 175°C로 예열된 오븐에서 약 10분간 노릇하게 구워 준비한다.
2. 화이트 베이스에 구운 아몬드를 넣고 믹서로 갈아준다.
3. 면보를 이용해 아몬드 고형분을 모두 걸러낸다.

- 걸러내고 난 후의 무게는 890g보다 적으므로, 부족한 중량은 화이트 베이스를 추가해 890g으로 맞춰 사용한다.

4. 냄비에 **3**의 2/3, 페퍼민트 잎을 넣고 75°C로 가열한다.
5. 75°C에 도달하면 뚜껑을 덮어 15~20분간 향을 우려낸 후 잎을 걸러낸다.
6. 비커에 **5**와 남은 **3**의 1/3, 생크림, 설탕, 민트 페이스트, 라임즙을 넣는다.
7. 핸드블렌더로 믹싱한다.
8. 제조기에 넣고 냉각 교반한다.

1. Roast almonds for 10 minutes until golden brown in an oven preheated to 175°C.
2. Add the almonds to white base and mix with a hand blender.
3. Strain all the solids using a cheesecloth.

- The sieved weight is less than 890 grams. Add more milk to adjust to 890 grams.

4. Add 2/3 of **3** and mint leaves in a pot and heat to 75°C.
5. When it reaches 75°C, close the lid to infuse for about 15~20 minutes and strain the leaves.
6. Add **5**, remaining 1/3 of **3**, cream, sugar, mint paste, and lime juice in a container.
7. Mix using a hand blender.
8. Pour into the machine and cold-churn the mixture.

싱글 레시피
Single recipe

Ingredients		Quantity
구운 아몬드	Roasted almonds	180g
우유	Milk	590g
생크림	Cream (38% fat)	130g
물엿	Glucose syrup	30g
페퍼민트 잎	Peppermint leaves	적당량 QS
탈지분유	Skim milk powder	50g
설탕	Sugar	140g
함수결정포도당	Dextrose	20g
복합안정제	Stabilizer	5g
라임즙	Lime juice	30g
민트 페이스트	Mint paste	5g
TOTAL		1,000g

● 페피민트 잎의 양은 기호에 맞게 조절한다.

● Adjust the amount of peppermint leaves to taste.

1. 아몬드는 175°C로 예열된 오븐에서 약 10분간 노릇하게 구워 준비한다.
2. 우유에 구운 아몬드를 넣고 믹서로 갈아준다.
3. 면보를 이용해 아몬드 고형분을 모두 걸러낸다.

● 걸러내고 난 후의 무게는 590g보다 적으므로, 부족한 중량은 우유를 추가해 590g으로 맞춰 사용한다.

4. 냄비에 3과 생크림, 물엿, 페퍼민트 잎을 넣고 40°C로 가열한다.
5. 믹싱볼에 탈지분유, 설탕, 함수결정포도당, 복합안정제를 계량한 후 스패출러로 혼합한다.
6. 5를 4에 넣고 65~85°C로 가열한다.
7. 4°C로 냉각한다.
8. 0~12시간 숙성한다.
9. 페퍼민트 잎을 걸러내고 라임즙과 민트 페이스트를 넣고 섞는다.
10. 제조기에 넣고 냉각 교반한다.

1. Roast almonds for about 10 minutes until golden brown in an oven preheated to 175°C.
2. Add the almonds to the milk and mix with a hand blender.
3. Strain all the solids using a cheesecloth.

● The sieved weight is less than 590 grams. Add more milk to adjust to 590 grams.

4. Heat 3, cream, glucose syrup, and mint leave in a pot to 40°C.
5. Measure skim milk powder, sugar, dextrose, and stabilizer to combine.
6. Mix 5 with 4 and heat to 65~85°C.
7. Cool to 4°C.
8. Let rest for 0~12 hours.
9. Strain the mint leaves and mix with mint paste and lime juice.
10. Pour into the machine and cold-churn the mixture.

당 Sugar	지방 Fat	무지유 고형분 Non-fat milk solids	기타 고형분 Other solids	총 고형분 Total solids	수분 Water
18.7%	7.1%	10.5%	0.8%	37.1%	62.9%

Memo.

14 Ispahan Gelato

Hot Infusion

이스파한 젤라또

이스파한은 프랑스의 디저트 셰프인 '피에르 에르메Pierre Herme'가 장미, 라즈베리, 리치를 사용해 만든 맛으로 전 세계 수많은 셰프들이 이 이스파한을 오마주한 디저트를 만들고 있다. 여기에서는 이스파한을 젤라또로 표현하기 위해 장미 꽃잎을 인퓨징하고, 라즈베리와 리치 포셰를 곁들여 식감도 더해보았다.

Ispahan is a flavor created by Pierre Herme, a French dessert chef, using roses, raspberries, and lychees, and numerous chefs around the world are making desserts that pay homage to Ispahan. I infused rose petals to express Ispahan as gelato and added texture with raspberries and lychee pochée.

베이스를 사용한 레시피
Recipe using the base

Ingredients		Quantity
화이트 베이스	White base	910g
장미꽃잎	Rose petals	10g
생크림	Cream (38% fat)	40g
함수결정포도당	Dextrose	20g
라즈베리&리치 시럽	Raspberry & lychee syrup	30g
라즈베리&리치 포셰	Raspberry & lychee pochée	적당량 QS
TOTAL		1,000g

● 라즈베리&리치 시럽과 포셰는 97p를 참고한다.

● Refer to p.97 for raspberry & lychee syrup and pochée.

1. 냄비에 화이트 베이스 2/3, 장미꽃잎을 넣고 75°C로 가열한다.
2. 온도에 도달하면 뚜껑을 덮어 15~20분간 향을 더 우려낸 후 장미꽃잎을 걸러낸다.
3. 비커에 2와 남은 화이트 베이스 1/3, 생크림, 함수결정포도당, 라즈베리&리치 시럽을 넣고 핸드블렌더로 믹싱한다.
4. 제조기에 넣고 냉각 교반한다.
5. 바트에 추출하며 라즈베리&리치 포셰를 섞어준다.

1. Heat 2/3 of white base and rose petals to 75°C.
2. When it reaches the temperature, close the lid and infuse the rose aroma for 15~20 minutes.
3. Combine 2, the remaining 1/3 of white base, cream, dextrose, and raspberry & lychee syrup; mix with a hand blender.
4. Pour into the machine and cold-churn the mixture.
5. Extract into a stainless-steel container and mix with raspberry & lychee pochée.

싱글 레시피
Single recipe

Ingredients		Quantity
우유	Milk	625g
물엿	Glucose syrup	20g
생크림	Cream (38% fat)	130g
장미꽃잎	Rose petals	10g
탈지분유	Skim milk powder	40g
설탕	Sugar	130g
함수결정포도당	Dextrose	30g
복합안정제	Stabilizer	5g
라즈베리&리치 시럽	Raspberry & lychee syrup	30g
라즈베리&리치 포셰	Raspberry & lychee pochée	적당량 QS
TOTAL		1,000g

1. 냄비에 우유, 물엿, 생크림, 장미꽃잎을 넣고 40°C로 가열한다.
2. 믹싱볼에 탈지분유, 설탕, 함수결정포도당, 복합안정제를 계량한 후 스패출러로 혼합한다.
3. 2를 1에 넣고 65~85°C로 가열한다.
4. 4°C로 냉각한다.
5. 라즈베리&리치 시럽을 넣고 0~12시간 숙성한다.
6. 장미꽃잎을 걸러낸다.
7. 제조기에 넣고 냉각 교반한다.
8. 바트에 추출하며 라즈베리&리치 포셰를 섞어준다.

1. Heat milk, glucose syrup, cream, and rose petals to 40°C.
2. Measure skim milk powder, sugar, dextrose, and stabilizer in a mixing bowl to combine.
3. Mix 2 with 1 and heat to 65~85°C.
4. Cool to 4°C.
5. Mix with raspberry & lychee syrup and let rest for 0~12 hours.
6. Strain the rose petals.
7. Pour into the machine and cold-churn the mixture.
8. Extract into a stainless-steel container and mix with raspberry & lychee pochée.

당 Sugar	지방 Fat	무지유 고형분 Non-fat milk solids	기타 고형분 Other solids	총 고형분 Total solids	수분 Water
18.7%	7.1%	9.7%	0.5%	36%	64%

15 Sei Profumi Gelato

Hot Infusion

세이 프로푸미 젤라또

이탈리아의 젤라또 학교인 '칼피지아니 젤라또 유니버시티**Carpigiani Gelato University**'에서 인퓨징 젤라또 재료 사용의 이해를 높이기 위해 항상 설명하는 맛이다. 6가지 향신료가 사용되지만 어느 하나 튀지 않고 조화롭게 느껴지는 이국적인 향이 매력적이다.

It is a flavor that is always taught to increase understanding of infusing gelato ingredients at 'Carpigiani Gelato University, an Italian gelato school. Six spices are used, but the exotic aroma doesn't stand out, and being in harmony is attractive.

베이스를 사용한 레시피
Recipe using the base

Ingredients		Quantity
옐로우 베이스	Yellow base	1,000g
바닐라빈	1/2 Vanilla bean	1/2개
시나몬 스틱	1 Cinnamon stick	1개
원두	3 Coffee beans	3알
정향	1 Clove	1알
레몬제스트	Zest of one lemon	레몬 1개 분량
오렌지제스트	Zest of half orange	오렌지 1/2개 분량
TOTAL		1,000g

1. 냄비에 옐로우 베이스 2/3와 바닐라빈, 시나몬 스틱, 으깬 원두, 정향, 레몬제스트, 오렌지제스트를 넣고 75°C로 가열한다.
2. 온도에 도달하면 뚜껑을 덮어 15~20분간 향을 더 우려낸 후 면보에 걸러낸다.
3. 비커에 2와 남은 옐로우 베이스 1/3을 넣고 핸드블렌더로 믹싱한다.
4. 제조기에 넣고 냉각 교반한다.

1. Combine 2/3 of yellow base, vanilla bean, cinnamon stick, crushed coffee beans, clove, lemon zest, and orange zest; heat to 75°C.
2. When it reaches the temperature, close the lid to infuse for 15~20 minutes and strain using a cheesecloth.
3. Add 2 and the remaining 1/3 of the yellow base and mix with a hand blender.
4. Pour into the machine and cold-churn the mixture.

싱글 레시피
Single recipe

Ingredients		Quantity
우유	Milk	610g
물엿	Glucose syrup	30g
생크림	Cream (38% fat)	95g
노른자	Egg yolks	60g
바닐라빈	1/2 Vanilla bean	1/2개
시나몬 스틱	1 Cinnamon stick	1개
원두	3 Coffee beans	3알
정향	1 Clove	1알
레몬제스트	Zest of one lemon	레몬 1개 분량
오렌지제스트	Zest of half orange	오렌지 1/2개 분량
탈지분유	Skim milk powder	25g
설탕	Sugar	135g
함수결정포도당	Dextrose	40g
복합안정제	Stabilizer	5g
TOTAL		1,000g

1. 냄비에 우유, 물엿, 생크림, 노른자, 바닐라빈, 시나몬 스틱, 으깬 원두, 정향, 레몬제스트, 오렌지제스트를 넣고 40°C로 가열한다.
2. 믹싱볼에 탈지분유, 설탕, 함수결정포도당, 복합안정제를 계량한 후 스패출러로 혼합한다.
3. 2를 1에 넣고 65~85°C로 가열한다.
4. 4°C로 냉각한다.
5. 0~12시간 숙성한다.
6. 인퓨징 재료(바닐라빈, 시나몬 스틱, 으깬 원두, 정향, 레몬제스트, 오렌지제스트)를 걸러낸다.
7. 제조기에 넣고 냉각 교반한다.

1. Combine milk, glucose syrup, cream, egg yolks, vanilla bean, cinnamon stick, crushed coffee beans, clove, lemon zest, orange zest; heat to 40°C.
2. Measure skim milk powder, sugar, dextrose, and stabilizer in a mixing bowl to combine.
3. Mix **2** with **1** and heat to 65~85°C.
4. Cool to 4°C.
5. Let rest for 0~12 hours.
6. Strain the infused ingredients (vanilla bean, cinnamon stick, crushed coffee beans, clove, lemon zest, orange zest).
7. Pour into the machine and cold-churn the mixture.

당 Sugar	지방 Fat	무지유 고형분 Non-fat milk solids	기타 고형분 Other solids	총 고형분 Total solids	수분 Water
19.6%	7.6%	8.1%	1.6%	36.8%	63.2%

Memo.

16 White Coco Gelato

Hot Infusion

화이트 코코 젤라또

이탈리아 교수님이 '나는 우울할 때 화이트초콜릿과 코코넛 젤라또를 함께 먹으면 기분이 좋아진다.'는 이야기를 듣고 만들어본 메뉴다. 교수님의 이야기를 듣고 나 역시 기분이 좋지 않을 때 즐겨 먹었던 조합이다. 달콤하면서도 부드러운 화이트초콜릿과 오독오독 씹히는 고소한 코코넛은 참 잘 어울리는 조합 중 하나다.

This is a menu I made after hearing from an Italian professor at a gelato school that when he is depressed, having white chocolate and coconut gelato together makes him feel better. It's a combination I also enjoyed when I was not in a good mood after listening to the story. Sweet yet soft white chocolate and nutty coconut with crunchy bite are among the best.

베이스를 사용한 레시피
Recipe using the base

Ingredients		Weight
구운 코코넛	Roasted coconut	55g
화이트 베이스	White base	550g
우유	Milk	100g
화이트초콜릿	White chocolate	100g
코코넛 퓌레	Coconut purée	250g
TOTAL		1,000g

1. 냄비에 구운 코코넛, 화이트 베이스 2/3, 우유를 넣고 75°C로 가열한다.
2. 온도에 도달하면 화이트초콜릿을 넣고 섞어준 후 뚜껑을 덮어 15~20분간 향을 더 우려낸다.
3. 2를 4°C로 냉각한다.
4. 구운 코코넛을 걸러낸다.

• 걸러내고 난 후의 무게는 550g보다 적으므로, 부족한 중량은 화이트 베이스를 추가해 550g으로 맞춰 사용한다.

5. 코코넛 퓌레를 넣고 핸드블렌더로 믹싱한다.
6. 제조기에 넣고 냉각 교반한다.

1. Heat roasted coconut, 2/3 of white base, and milk to 75°C.
2. When it reaches the temperature, mix with white chocolate, close the lid; infuse for 15~20 minutes.
3. Cool 2 to 4°C.
4. Strain the roasted coconut.

• The sieved weight is less than 550 grams. Add more white base to adjust to 550 grams.

5. Add coconut purée and mix using a hand blender.
6. Pour into the machine and cold-churn the mixture.

싱글 레시피
Single recipe

Ingredients		Weight
구운 코코넛	Roasted coconut	55g
우유	Milk	500g
물엿	Glucose syrup	10g
탈지분유	Skim milk powder	50g
설탕	Sugar	65g
함수결정포도당	Dextrose	20g
복합안정제	Stabilizer	5g
화이트초콜릿	White chocolate	100g
코코넛 퓌레	Coconut purée	250g
TOTAL		1,000g

1. 냄비에 구운 코코넛, 우유, 물엿을 넣고 40°C로 가열한다.
2. 믹싱볼에 탈지분유, 설탕, 함수결정포도당, 복합안정제를 계량한 후 스패출러로 혼합한다.
3. 2를 1에 넣고 65~85°C로 가열한다.
4. 온도에 도달하면 화이트초콜릿을 넣고 섞어준다.
5. 4°C로 냉각한다.
6. 코코넛 퓌레를 넣고 핸드블렌더로 믹싱한 후 0~12시간 숙성한다.
7. 구운 코코넛을 걸러낸다.

• 걸러내고 난 후의 무게는 500g보다 적으므로, 부족한 중량은 우유를 추가해 500g으로 맞춰 사용한다.

8. 제조기에 넣고 냉각 교반한다.

1. Heat roasted coconut, milk, and glucose syrup in a pot to 40°C.
2. Measure skim milk powder, sugar, dextrose, and stabilizer in a mixing bowl to combine.
3. Mix 2 with 1 and heat to 65~85°C.
4. When it reaches the temperature, add white chocolate to mix.
5. Cool to 4°C.
6. Add coconut purée, mix using a hand blender, and let rest for 0~12 hours.
7. Strain the roasted coconut.

• The sieved weight is less than 500 grams. Add more milk to adjust to 500 grams.

8. Pour into the machine and cold-churn the mixture.

당 Sugar	지방 Fat	무지유 고형분 Non-fat milk solids	기타 고형분 Other solids	총 고형분 Total solids	수분 Water
18.6%	11.7%	9.6%	1%	41%	59%

17

Caffé Bianco Gelato

Cold Infusion

카페 비안코 젤라또

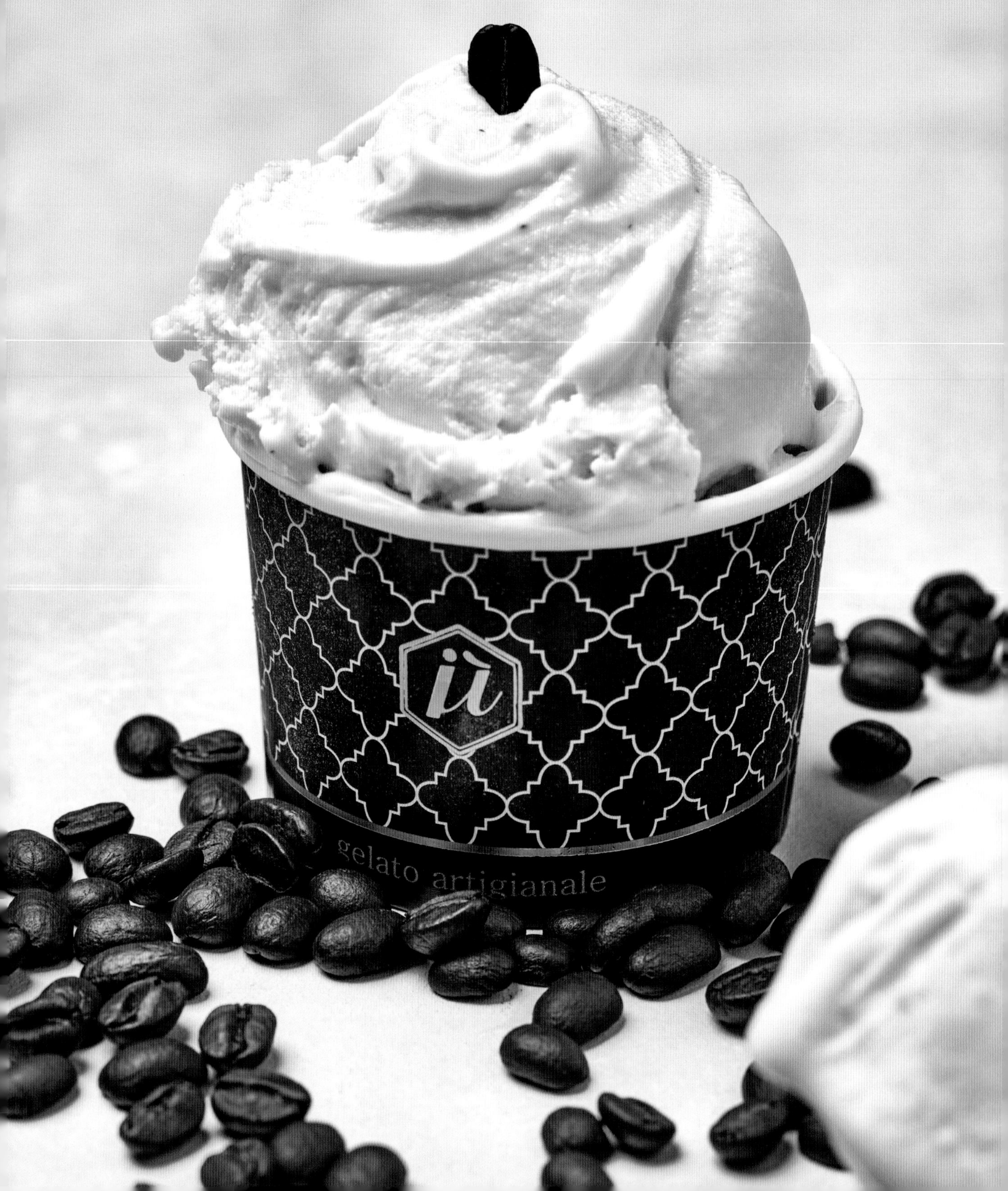

카페 비안코는 이탈리아어로 '하얀 커피'라는 뜻이다. 뜻 그대로 색은 하얗지만 먹었을 때 코끝에서 스쳐 지나가는 듯한 은은한 커피 향이 매력적인 젤라또이다.

Caffé Bianco means 'white coffee' in Italian. The color is white, as it is meant to be, but when consumed, this attractive gelato will give a subtle coffee aroma as if it passes the tip of your nose.

베이스를 사용한 레시피
Recipe using the base

Ingredients		Quantity
원두	Coffee beans	80g
화이트 베이스	White base	870g
생크림	Cream (38% fat)	85g
설탕	Sugar	45g
TOTAL		1,000g

- 원두의 양은 기호에 맞게 조절한다.
- 브라운 커피를 원할 경우, 아몬드 & 민트 젤라또(162p)와 같은 방법으로 Hot Infusion 기법으로 제조한다.

- Adjust the quantity of coffee beans to taste.
- If you want brown-colored coffee, use the hot infusion method as with the Almond & Mint Gelatto (p.162).

1. 비커에 원두를 넣고 도구를 이용해 절반 정도의 크기로 부순다.
2. 1에 화이트 베이스를 넣고 냉장고에서 12~24시간 동안 원두의 향을 우려낸 후 원두를 걸러낸다.
   - 걸러내고 난 후의 무게는 870g보다 적으므로, 부족한 중량은 화이트 베이스를 추가해 870g으로 맞춰 사용한다.
3. 비커에 2와 생크림, 설탕을 넣는다.
4. 3을 핸드블렌더로 믹싱한다.
5. 제조기에 넣고 냉각 교반한다.

1. Put coffee beans in a container and crush them into about half the size.
2. Add white base into 1 and infuse the aroma of the beans in the refrigerator for 12~24 hours.
   - The sieved weight is less than 870 grams. Add more white base to adjust to 870 grams.
3. Add 2, cream, and sugar in a beaker.
4. Mix 3 with a hand blender.
5. Pour into the machine and cold-churn the mixture.

싱글 레시피
Single recipe

Ingredients		Quantity
원두	Coffee beans	80g
우유	Milk	585g
물엿	Glucose syrup	30g
생크림	Cream (38% fat)	170g
탈지분유	Skim milk powder	40g
설탕	Sugar	140g
함수결정포도당	Dextrose	30g
복합안정제	Stabilizer	5g
TOTAL		1,000g

1. 비커에 원두를 넣고 도구를 이용해 절반 정도의 크기로 부순다.
2. 냄비에 우유, 물엿, 생크림을 넣고 40°C로 가열한다.
3. 믹싱볼에 탈지분유, 설탕, 함수결정포도당, 복합안정제를 계량한 후 스패출러로 혼합한다.
4. 3을 2에 넣고 65~85°C로 가열한다.
5. 4°C로 냉각한다.
6. 5에 1을 넣고 냉장고에서 12~24시간 숙성한 후 원두를 걸러낸다.
   - 걸러내고 난 후의 무게는 585g보다 적으므로, 부족한 중량은 우유를 추가해 585g으로 맞춰 사용한다.
7. 제조기에 넣고 냉각 교반한다.

1. Put coffee beans in a container and crush them into about half the size.
2. Heat milk, glucose syrup, and cream in a pot to 40°C.
3. In a mixing bowl, measure skim milk powder, sugar, glucose, and stabilizer and mix with a spatula.
4. Mix 3 with 2 and heat to 65~85°C.
5. Cool to 4°C.
6. Mix 5 with 1 and refrigerate for 12~24 hours, then sieve the beans.
   - The sieved weight is less than 585 grams. Add more milk to adjust to 585 grams.
7. Pour into the machine and cold-churn the mixture.

당 Sugar	지방 Fat	무지유 고형분 Non-fat milk solids	기타 고형분 Other solids	총 고형분 Total solids	수분 Water
19.2%	8.5%	9.7%	0.5%	37.9%	62.1%

18 Vegan Oat Chai Gelato

Vegan

비건 오트 차이 젤라또

귀리와 물을 혼합해 만든 대체 우유인 오트 밀크에 다양한 향신료가 혼합된 차이티를 인퓨징해 만든 젤라또. 오트 밀크에서 느껴지는 특유의 고소함과 차이티의 스파이시한 맛과 향이 조화롭게 어우러진다.

This gelato is made by infusing oat milk, an alternative milk made with oats, and chai tea with various spices. The unique mild, earthy note of oat milk and the spicy taste and aroma of chai tea harmonizes well.

싱글 레시피
Single recipe

Ingredients		Quantity
오트밀크	Oat milk	725g
식물성오일	Vegetable oil	50g
물엿	Glucose syrup	50g
차이티	Chai tea leaves	10g
설탕	Sugar	40g
트레할로스	Trehalose	95g
함수결정포도당	Dextrose	20g
이눌린	Inulin	15g
복합안정제	Stabilizer	5g
TOTAL		1,000g

1. 냄비에 오트밀크, 식물성오일, 물엿, 차이티를 넣고 40°C로 가열한다.

- 오트밀크는 오틀리(OATLY)를 사용했고, 차이티는 프라나 차이(PRANA CHAI)를 사용했다. 식물성오일은 향이 강하지 않은 것(예: 해바라기씨유, 포도씨유)이라면 모두 사용 가능하다.

2. 믹싱볼에 설탕, 트레할로스, 함수결정포도당, 이눌린, 복합안정제를계량한 후 스패출러로 혼합한다.
3. 2를 1에 넣고 65~85°C로 가열한다.
4. 4°C로 냉각한다.
5. 0~12시간 숙성한다.
6. 차이티를 걸러낸다.
7. 제조기에 넣고 냉각 교반한다.

1. Heat oat milk, oil, glucose syrup, and chai tea in a pot to 40°C.

- I used Oatly for oat milk and Prana Chai for chai tea. Any vegetable oil works as long as it does not have a strong odor (e.g., sunflower seed oil, grapeseed oil).

2. Measure sugar, trehalose, dextrose, inulin, and stabilizer in a mixing bowl to combine.
3. Mix 2 with 1 and heat to 65~85°C.
4. Cool to 4°C.
5. Let rest for 0~12 hours.
6. Strain the tea leaves.
7. Pour into the machine and cold-churn the mixture.

당 Sugar	지방 Fat	무지유 고형분 Non-fat milk solids	기타 고형분 Other solids	총 고형분 Total solids	수분 Water
22.4%	6%	-	5%	33.4%	66.6%

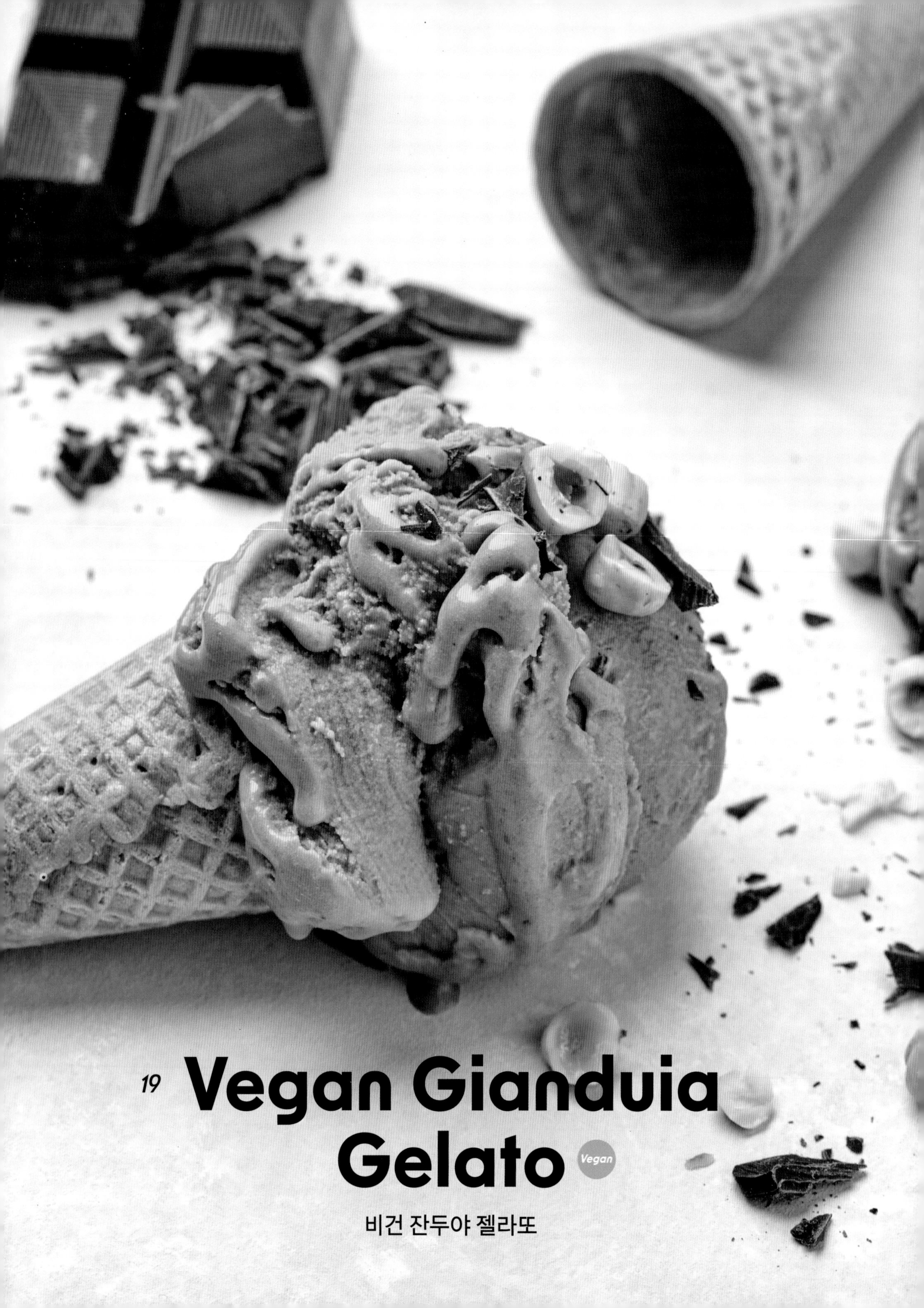

19 Vegan Gianduia Gelato

Vegan

비건 잔두야 젤라또

피에몬테산 헤이즐넛이 주는 묵직한 고소함과 다크초콜릿의 쌉싸래한 맛을 담았다. 유제품이 들어가지 않는 비건 젤라또라 사용하는 재료 본연의 맛을 그대로 느낄 수 있다.

It portrays the rich nutty taste of hazelnuts from Piedmont and the bitterness of dark chocolate. Because it's a vegan gelato that does not contain dairy products, you can taste the natural flavor of the ingredients.

싱글 레시피
Single recipe

Ingredients		Quantity
아몬드밀크	Almond milk	660g
물엿	Glucose syrup	30g
설탕	Sugar	40g
트레할로스	Trehalose	95g
함수결정포도당	Dextrose	40g
복합안정제	Stabilizer	5g
다크초콜릿	Dark Chocolate	50g
헤이즐넛	Hazelnuts	80g
TOTAL		1,000g

1. 냄비에 아몬드밀크, 물엿을 넣고 40°C로 가열한다.

● 아몬드밀크는 아몬드브리즈 언스위트(Almond Breeze unsweeted)를 사용했다.

2. 믹싱볼에 설탕, 트레할로스, 함수결정포도당, 복합안정제를 계량한 후 스패출러로 혼합한다.
3. 2를 1에 넣고 65~85°C로 가열한다.
4. 3에 다크초콜릿을 넣고 녹인다.
5. 4°C로 냉각한다.
6. 헤이즐넛을 넣고 핸드믹서로 믹싱한다.

● 헤이즐넛의 입자감이 느껴지는 것을 원한다면 핸드블렌더로 작업한다. 그렇지 않다면 푸드프로세서나 콘칭기로 페이스트화시켜 사용한다. 시판 헤이즐넛 페이스트(100%)를 사용할 수도 있으며, 이 경우 더 진한 맛으로 완성된다.

7. 0~12시간 숙성한다.
8. 제조기에 넣고 냉각 교반한다.

1. Heat almond milk and glucose syrup in a pot to 40°C.

● I used Almond Breeze unsweetened milk.

2. Measure sugar, trehalose, dextrose, and stabilizer in a mixing bowl and combine with a spatula.
3. Mix 2 with 1 and heat to 65~85°C.
4. Melt the dark chocolate and mix with 3.
5. Cool to 4°C.
6. Add the hazelnuts and mix with a hand blender.

● If you don't want the grainy texture of hazelnuts, grind with a hand blender. If not, make a paste using a food processor or conching machine. You can also use commercially available hazelnut paste (100%), which will give a richer taste.

7. Let rest for 0~12 hours.
8. Pour into the machine and cold-churn the mixture.

당 Sugar	지방 Fat	무지유 고형분 Non-fat milk solids	기타 고형분 Other solids	총 고형분 Total solids	수분 Water
21.1%	7.8%	-	6%	34.9%	65.1%

20 Berry Wine Gelato

Alcohol

베리 와인 젤라또

레드와인이 가진 탄닌감을 중화시켜주고, 유지방과 섞였을 때 감소되는 프루티한 과일의 향을 끌어올리기 위해 트리플베리를 곁들여 만든 메뉴다.

It is a menu made with triple berries to neutralize the tannin of red wine and to enhance the fruity flavor that diminishes when mixed with milk fat.

싱글 레시피
Single recipe

Ingredients		Weight
우유	Milk	335g
생크림	Cream (38% fat)	80g
탈지분유	Skim milk powder	60g
설탕	Sugar	90g
말토덱스트린	Maltodextrin	25g
이눌린	Inulin	15g
복합안정제	Stabilizer	5g
트리플 베리 퓌레	Triple Berry purée	200g
레드와인	Red wine	190g
TOTAL		1,000g

● Contains 2.5% alcohol

1. 냄비에 우유, 생크림을 넣고 40°C로 가열한다.
2. 믹싱볼에 탈지분유, 설탕, 말토덱스트린, 이눌린, 복합안정제를 계량한 후 스패출러로 혼합한다.
3. 2를 1에 넣고 65~85°C로 가열한다.
4. 4°C로 냉각한다.
5. 트리플 베리 퓌레를 넣고 핸드블렌더로 믹싱한다.
6. 0~12시간 숙성한다.
7. 제조기에 넣고 냉각 교반한다.

● 40% 정도 냉각되었을 때 레드와인을 넣는다.

1. Heat milk and cream in a pot to 40°C.
2. Measure skim milk powder, sugar, maltodextrin, inulin, and stabilizer in a mixing bowl and combine with a spatula.
3. Mix 2 with 1 and heat to 65~85°C.
4. Cool to 4°C.
5. Add the purée and mix with a hand blender.
6. Let rest for 0~12 hours.
7. Pour into the machine and cold-churn the mixture.

● Add the red wine when the mixture is about 40% chilled.

당 Sugar	지방 Fat	무지유 고형분 Non-fat milk solids	기타 고형분 Other solids	총 고형분 Total solids	수분 Water
17.6%	4.5%	9%	2%	32.6%	67.4%

21 Chestnut Marnier Gelato

Alcohol

체스트넛 마르니에르 젤라또

'그랑 마르니에르Grand Marnier'는 향긋한 오렌지 향과 오크 향이 함께 느껴지는 프랑스의 브랜디이다. 자칫 강하게 느껴질 수 있는 알코올 향을 체스트넛&바닐라 퓌레를 더해 부드럽고 달콤하게 잡아주었다.

'Grand Marnier' is a French brandy with fragrant orange with oak aroma. The taste of alcohol, which can be quite strong, was softened and sweetened by adding chestnut and vanilla puree.

싱글 레시피
Single recipe

Ingredients		Quantity
우유	Milk	400g
생크림	Cream (38% fat)	140g
탈지분유	Skim milk powder	55g
설탕	Sugar	55g
말토덱스트린	Maltodextrin	20g
이눌린	Inulin	15g
복합안정제	Stabilizer	5g
체스트넛 & 바닐라 퓌레	Chestnut & Vanilla purée	250g
그랑 마르니에르	Grand Marnier	60g
TOTAL		1,000g

● Contains 2.5% alcohol

1. 냄비에 우유, 생크림을 넣고 40°C로 가열한다.
2. 믹싱볼에 탈지분유, 설탕, 말토덱스트린, 이눌린, 복합안정제를 계량한 후 스패출러로 혼합한다.
3. 2를 1에 넣고 65~85°C로 가열한다.
4. 4°C로 냉각한다.
5. 체스트넛&바닐라 퓌레를 넣고 핸드블렌더로 믹싱한다.
6. 0-12시간 숙성한다.
7. 제조기에 넣고 냉각 교반한다.

● 40% 정도 냉각되었을 때 그랑 마르니에르를 넣는다.

1. Heat milk and cream in a pot to 40°C.
2. Measure skim milk powder, sugar, maltodextrin, inulin, and stabilizer in a mixing bowl and combine with a spatula.
3. Mix 2 with 1 and heat to 65~85°C.
4. Cool to 4°C.
5. Add the purée and mix with a hand blender.
6. Let rest for 0~12 hours.
7. Pour into the machine and cold-churn the mixture.

● Add Grand Marnier when the mixture is about 40% chilled.

당 Sugar	지방 Fat	무지유 고형분 Non-fat milk solids	기타 고형분 Other solids	총 고형분 Total solids	수분 Water
18%	6.8%	9.5%	2%	36.3%	63.7%

22 # Gorgonzola & Pistachio Gelato

Savory

고르곤졸라 & 피스타치오 젤라또

2016년 '라보라또리오 일쀼Laboratorio il più'를 운영할 때 우연히 이탈리아의 젤라띠에레가 만든 고르곤졸라&피스타치오 젤라또 사진을 보고 어떤 맛일지, 과연 잘 어울리는 조합일지 의문이 들어 직접 만들어본 나의 첫 세이보리 젤라또이다. 고르곤졸라 특유의 독특한 향을 피스타치오의 진한 지방감으로 중화시켜 일반적이지는 않지만 중독성 강한 맛이다.

When I was running 'Laboratorio il più' in 2016, I happened to see a picture of Gorgonzola & Pistachio Gelato made by a gelatiere in Italy, and I wondered what it would taste like and whether it would go well together, so this became the first savory gelato I made. Gorgonzola's unique flavor is neutralized by pistachio's rich fattiness, making it an unusual but addictive taste.

캔디드 피스타치오*
Candied pistachios

Ingredients		Quantity
설탕	Sugar	70g
물	Water	30g
피스타치오	Pistachios	100g
TOTAL		200g

1. 팬에 설탕과 물을 넣고 투명한 시럽 상태가 될 때까지 가열하면서 녹인다.
2. 피스타치오를 넣고 하얗게 코팅이 될 때까지 저어가며 설탕 시럽을 입힌다.
3. 계속 저어가며 하얗게 코팅된 설탕이 다시 녹아 캐러멜라이징이 될 때까지 가열한다.

● 기호에 따라 다른 견과류로 대체해 제조할 수 있다.

1. Put sugar and water in a pot and heat until the sugar dissolves and turns into a translucent syrup.
2. Add the pistachios and stir to coat the nuts until the sugar recrystallizes and turns white.
3. Put the pan back on the heat and continue to stir until the sugar melts and caramelizes.

● Nuts can be replaced to taste.

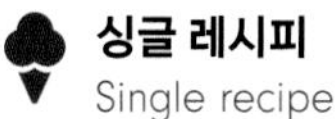

싱글 레시피
Single recipe

Ingredients		Quantity
우유	Milk	460g
물	Water	180g
올리브오일	Olive oil	20g
버터	Butter	20g
물엿	Glucose syrup	35g
탈지분유	Skim milk powder	50g
트레할로스	Trehalose	130g
함수결정포도당	Dextrose	30g
복합안정제	Stabilizer	5g
소금	Salt	2g
고르곤졸라	Gorgonzola	30g
피스타치오 페이스트	Pistachio paste	40g
캔디드 피스타치오*	Candied pistachios*	적당량 QS
TOTAL		1,000g

1. 냄비에 우유, 물, 올리브오일, 버터, 물엿을 넣고 40°C로 가열한다.
2. 믹싱볼에 탈지분유, 트레할로스, 함수결정포도당, 복합안정제, 소금을 계량한 후 스패출러로 혼합한다.
3. 2를 1에 넣고 65~85°C로 가열한다.
4. 4°C로 냉각한다.
5. 고르곤졸라와 피스타치오 페이스트를 넣고 핸드블렌더로 믹싱한다.
6. 0~12시간 숙성한다.
7. 제조기에 넣고 냉각 교반한다.
8. 바트에 추출하며 캔디드 피스타치오를 섞어준다.

1. Heat milk, water, olive oil, butter, and glucose syrup in a pot to 40°C.
2. Measure skim milk powder, trehalose, dextrose, and stabilizer in a mixing bowl and combine with a spatula.
3. Mix 2 with 1 and heat to 65~85°C.
4. Cool to 4°C.
5. Add gorgonzola and pistachio paste; mix with a hand blender.
6. Let rest for 0~12 hours.
7. Pour into the machine and cold-churn the mixture.
8. Extract into a stainless-steel container and mix with candied pistachios.

당 Sugar	지방 Fat	무지유 고형분 Non-fat milk solids	기타 고형분 Other solids	총 고형분 Total solids	수분 Water
18.5%	8.3%	9.2%	2.3%	38.4%	61.6%

23 Potato & Bacon Gelato

Savory

감자 & 베이컨 젤라또

젤라떼리아에서 세이보리 젤라또를 콘이나 컵으로 판매하기에는 쉽지 않을 것이다. 하지만 이 젤라또는 포슬포슬한 감자와 메이플 시럽을 입힌 베이컨의 조합으로 젤라떼리아에서도 충분히 판매할 수 있는 달고 짠맛의 최고 조합이다. 한번 먹어본 사람들은 반드시 찾게 되는 맛으로 창업반 수업에서도 소개하는 인기 메뉴이다.

It is challenging for gelateria to sell savory gelato in cones or cups. However, this gelato is made with fluffy potato and maple syrup-coated bacon, which presents the best sweet and salty combination that can be sold in gelateria. It's a popular menu I introduce in entrepreneurship class as it is a taste customers will return for.

캔디드 베이컨*
Candied bacon

Ingredients		Quantity
베이컨	Bacon	적당량 QS
메이플 시럽	Maple syrup	적당량 QS
TOTAL		200g

1. 팬에 베이컨을 초벌로 구운 후 키친타월에 올려 기름기를 제거한다.
2. 기름기를 제거한 베이컨에 메이플 시럽을 두 번 바른다.
3. 다시 팬에 올려 뒤집어가며 타지 않게 굽는다.
4. 적당한 크기로 자른다.

1. Par-cook the bacon in a pan and place on a kitchen towel to remove excess fat.
2. Brush the degreased bacon twice with maple syrup.
3. Put them back in the pan and cook evenly without scorching.
4. Cut to moderate size.

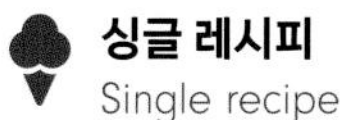

싱글 레시피
Single recipe

Ingredients		Quantity
우유	Milk	360g
물	Water	140g
올리브오일	Olive oil	20g
버터	Butter	25g
물엿	Glucose syrup	35g
탈지분유	Skim milk powder	50g
트레할로스	Trehalose	130g
함수결정포도당	Dextrose	40g
소금	Salt	2g
구운 감자	Baked potatoes	200g
캔디드 베이컨*	Candied bacon*	적당량 QS
TOTAL		1,000g

1. 냄비에 우유, 물, 올리브오일, 버터, 물엿을 넣고 40°C로 가열한다.
2. 믹싱볼에 탈지분유, 트레할로스, 함수결정포도당, 소금을 계량한 후 스패출러로 혼합한다.
3. 2를 1에 넣고 65~85°C로 가열한다..
4. 4°C로 냉각한다.
5. 구운 감자를 넣고 핸드블렌더로 믹싱한다.
6. 0~12시간 숙성한다.
7. 제조기에 넣고 냉각 교반한다.
8. 바트에 추출하며 캔디드 베이컨을 섞어준다.

1. Heat milk, water, olive oil, butter, and glucose syrup in a pot to 40°C.
2. Measure skim milk powder, trehalose, dextrose, and salt in a mixing bowl and combine with a spatula.
3. Mix 2 with 1 and heat to 65~85°C.
4. Cool to 4°C.
5. Add baked potatoes and mix with a hand blender.
6. Let rest for 0~12 hours.
7. Pour into the machine and cold-churn the mixture.
8. Extract into a stainless-steel container and mix with candied bacon.

당 Sugar	지방 Fat	무지유 고형분 Non-fat milk solids	기타 고형분 Other solids	총 고형분 Total solids	수분 Water
19.5%	5%	8.2%	3.5%	36.2%	63.8%

24 Camembert & Ssamjang Gelato

Savory

까망베르 & 쌈장 젤라또

2022년 싱가포르에서 열린 '아시안 젤라또 컵' 대회에서 세이보리 젤라또 미션을 받고 선보인 메뉴다. 잎채소에 밥, 고기, 쌈장을 싸 먹는 한국의 식문화를 젤라또 방식으로 표현해보았다. 외국인 심사위원들이 쌈장을 너무 부담스럽게 느끼지 않도록 까망베르를 더했고, 쌈을 표현하기 위해 엔다이브 위에 잠봉으로 만든 리조또를 곁들였다.

This menu was introduced for the Savory Gelato mission at the Asian Gelato Cup held in Singapore in 2022. The Korean food culture of wrapping rice, meat, and ssamjang in leaf vegetables was expressed into a gelato. Camembert was added so that the judges would not feel that ssamjang is too strong, and risotto made with jambon was served on top of endive leaves to convey ssam (wrap).

잠봉 리조또*
Jambon risotto

Ingredients		Quantity
버터	Butter	5g
잠봉	Jambon	75g
화이트와인	White wine	적당량 QS
우유	Milk	150g
노른자	Egg yolks	20g
쌀	Rice	100g
소금	Salt	적당량 QS
후추	Pepper	적당량 QS
TOTAL		350g

1. 후라이팬에 버터와 잠봉을 넣고 볶는다.
2. 온도가 오르면 화이트와인으로 플람베한다.
3. 잠봉이 구워지면 우유, 노른자, 쌀을 넣고 졸인다.
4. 소금과 후추로 간을 한다.

1. Stir-fry jambon ham with butter in a frying pan.
2. When the temperature rises, flambé with white wine.
3. Once jambon ham is cooked, add milk, egg yolks, and rice to simmer.
4. Season with salt and pepper to taste.

싱글 레시피
Single recipe

Ingredients		Quantity
쌈장	Ssamjang	30g
마늘	Garlic	5g
우유	Milk	420g
물	Water	135g
생크림	Cream (38% fat)	50g
물엿	Glucose syrup	35g
탈지분유	Skim milk powder	45g
트레할로스	Trehalose	115g
함수결정포도당	Dextrose	25g
복합안정제	Stabilizer	5g
까망베르	Camembert	135g
잠봉 리조또*	Jambon risotto*	적당량 QS
TOTAL		1,000g

1. 냄비에 쌈장과 마늘을 넣고 마늘이 익을 때까지 볶는다.
2. 1에 우유, 물, 생크림, 물엿을 넣고 40°C로 가열한다.
3. 믹싱볼에 탈지분유, 트레할로스, 함수결정포도당, 복합안정제를 계량한 후 스패출러로 혼합한다.
4. 3을 2에 넣고 65~85°C로 가열한다.
5. 4°C로 냉각한다.
6. 0~12시간 숙성한다.
7. 체망에 받쳐 6을 걸러낸다.
8. 걸러진 7에 까망베르를 넣고 핸드블렌더로 믹싱한다.
9. 제조기에 넣고 냉각 교반한다.
10. 엔다이브 또는 잎채소에 잠봉 리조또와 젤라또를 함께 올려 먹는다.

1. Stir-fry ssamjang and garlic in a pot until the garlic is cooked.
2. Add milk, water, cream, and glucose syrup to 1 and heat to 40°C.
3. Measure skim milk powder, trehalose, dextrose, and stabilizer in a mixing bowl and combine with a spatula.
4. Mix 3 with 2 and heat to 65~85°C.
5. Cool to 4°C.
6. Let rest for 0~12 hours.
7. Strain 6 through a sieve.
8. Add camembert to 7 and mix with a hand blender.
9. Pour into the machine and cold-churn the mixture.
10. Enjoy Jambon risotto and gelato together on endives or leaf vegetables.

당 Sugar	지방 Fat	무지유 고형분 Non-fat milk solids	기타 고형분 Other solids	총 고형분 Total solids	수분 Water
17.5%	6.5%	9.4%	1%	34.4%	65.6%

기계로 만든 젤라또
Gelato made using a batch freezer

믹서(블렌더)를 이용해 만든 젤라또
Gelato made using a mixer (blender)

얼리면서 긁고 섞어 만든 젤라또
Gelato made by scraping and mixing while freezing

HOMEMADE GELATO RECIPE

: How to Make Gelato at Home Without a Machine

홈메이드 젤라또 레시피

: 집에서 기계 없이 젤라또를 만드는 방법

집에서 기계 없이 쉽고 간단하게 젤라또를 만드는 방법은 두 가지가 있다.

첫 번째 방법은 제시된 레시피와 동일하게 재료를 계량해 믹싱한 후 냉동 보관이 가능한 용기에 담고, 2~3시간마다 한 번씩 꺼내 포크나 숟가락으로 긁고 섞어주며 공기를 주입하는 것이다. 이 과정을 두 세 번 반복하면 얼음 입자가 작게 형성되어 단단하지 않은 질감으로 젤라또를 즐길 수 있다.

There are two easy and simple ways to make gelato at home without a machine.

The first method; weigh and mix the same way as the recipe provided, then place it in a container that can be stored frozen. Take it out every 2~3 hours, scrape it with a fork or spoon, and mix to incorporate air. When you repeat this process two or three times, the ice particles are formed small, and you can make a gelato that is not hard in texture.

첫 번째 방법보다 더 많은 공기 주입으로 젤라또를 부드럽게 만들 수 있는 방법은 믹서기(블렌더)를 사용하는 것이다. 제시된 레시피와 동일하게 재료를 계량한 후 얼음 틀에 믹스를 부어 얼린다. 그 다음 단단하게 언 믹스를 믹서기에 넣고 곱게 갈아준 후 냉동 보관이 가능한 용기에 담아 20분 정도 냉동 보관하면 완성이다.

You can make a more aerated and softer gelato than the first method by using a blender. Weigh the ingredients the same way as the recipe provided, pour the mixture into an ice cube mold, and freeze. Grind the frozen mixture finely in a blender, put it in a freezable container, and freeze it for about 20 minutes to finish.

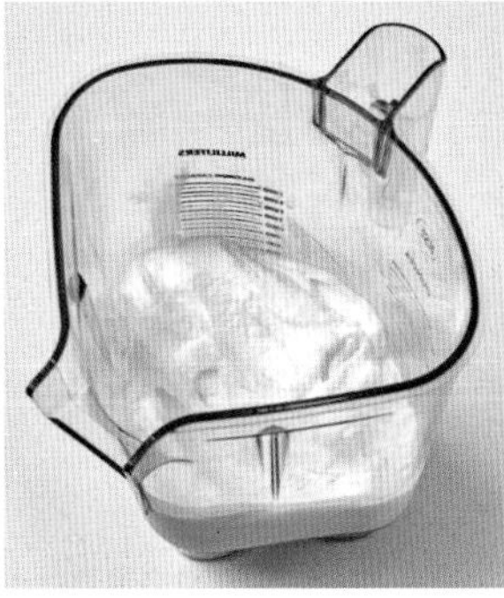
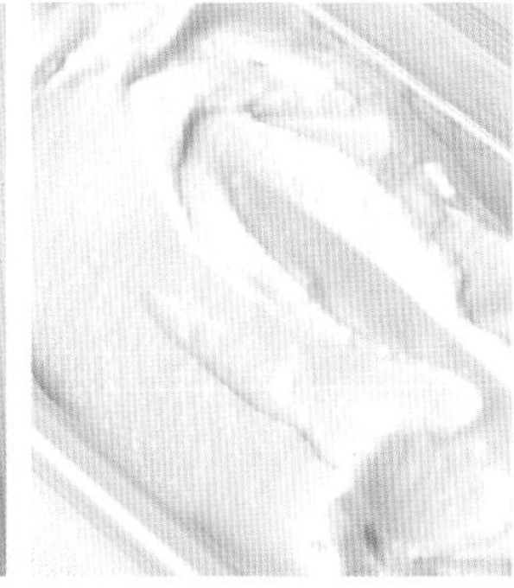

✦ 집에서 만들 때는 레시피 재료 중 마트에서 구입하기 어려운 '복합안정제'를 생략해도 된다. 하지만 젤라또의 찰진 질감은 낼 수 없다.

✦ You can omit the 'stabilizer' among the ingredients when making it at home since it can be challenging to find in supermarkets. However, you cannot make the dense and elastic texture of gelato.

Part 6.

Theory for Writing Sorbetto Recipes

소르베또 레시피를 작성하기 위한 이론

Note

- ✓ 레시피를 작성하는 방법은 여러 가지가 있지만, 그 중 가장 쉽게 결과값을 찾을 수 있는 Carpigiani Gelato University의 방법으로 설명하였다.
- ✓ 고형분에서 소수점은 1자리까지 봐주고, 중량에서 소수점은 올리거나 버렸다.

- ✓ There are many ways to write a recipe, but Carpigiani Gelato University's method was explained, which is the easiest to find the result value.
- ✓ The decimal point was counted to one digit, and the decimal point was raised or dropped for the weight.

1 Writing and Understanding Sorbetto Recipes

소르베또 레시피 작성과 이해

유제품이 들어가는 젤라또와는 다르게 소르베또는 유제품이 단 1%도 들어가지 않는다. 용어의 혼재로 소르베또**Sorbetto**와 셔벗**Sherbet**이 같은 콜드 디저트로 간주되지만 실제로는 비슷한 듯 다른 디저트이다.

소르베또는 물, 주재료, 당, 안정제로 구성되어 있어 시원하고 깔끔한 청량감이 느껴지는 것이 특징이다. 셔벗은 소르베또처럼 물이 들어가지만 보다 더 부드러운 질감을 부여하기 위해 우유나 탈지분유 등 소량의 유제품이 혼합된다. 그렇기 때문에 비건이나 유당불내증을 가진 고객에게 셔벗을 권해서는 안된다.

이탈리아 젤라떼리아에서는 셔벗을 찾아보기 힘들지만, 미국에서는 셔벗과 소르베또를 명확하게 구분지어 판매하는 것을 볼 수 있다.

레시피 작성의 가장 첫 번째 순서는 사용하는 재료들의 고형분을 정확하게 파악하는것이다. 그래야 균형 잡힌 소르베또를 만들 수 있다.

Unlike gelato, that contains dairy products, sorbetto does not contain even 1% of dairy ingredient. Due to the blended use of terms, sorbetto and sherbet are considered to be the same cold desserts, but they are indeed different dessert that seem similar.

Sorbetto is composed of water, a main ingredient, sugar, and a stabilizer, characterized by its cool and refreshing taste. Sherbet contains water like sorbetto but is mixed with a small number of dairy products, such as milk or skim milk powder to give it a softer texture. For this reason, sherbet should not be recommended to vegans or customers with lactose intolerance.

You can rarely find sherbet in Italian gelateria, but you can see that sherbet and sorbetto are clearly separated in sold in the United States.

The first step in creating a recipe is to determine the solid content of the ingredients used accurately. That way, you can make a good balanced sorbetto.

소르베또에 사용되는 재료들의 평균 고형분 함량

Average solids content of ingredients used in sorbetto

과일 Fruits	당 Sugar	지방 Fat	무지유 고형분 Non-fat milk solids	기타 고형분 Other solids	총 고형분 Total solids	수분 Water
바나나 Banana	14%	-	-	9%	23%	77%
수박 Watermelon	6%	-	-	5%	11%	89%
사과 Apple	8%	-	-	6%	14%	86%
파인애플 Pineapple	12%	-	-	9%	21%	79%
오렌지 Orange	8%	-	-	8%	16%	84%
자몽 Grapefruit	8%	-	-	6%	14%	86%
키위 Kiwi	7%	-	-	3%	10%	90%
망고 Mango	13%	-	-	8%	21%	79%
복숭아 Peach	9%	-	-	9%	18%	82%

과일 Fruits	당 Sugar	지방 Fat	무지유 고형분 Non-fat milk solids	기타 고형분 Other solids	총 고형분 Total solids	수분 Water
포도 Grape	20%	-	-	5%	25%	75%
딸기 Strawberry	8%	-	-	7%	15%	85%
레몬 Lemon	5%	-	-	8%	13%	87%
살구 Apricot	9%	-	-	9%	18%	82%
체리 Cherry	10%	-	-	6%	16%	84%
라즈베리 Raspberry	7%	-	-	7%	15%	85%
귤 Tangerine	8%	-	-	8%	16%	84%
멜론 Melon	8%	-	-	5%	13%	87%
패션프루트 Passion fruit	7%	-	-	4%	11%	89%
파파야 Papaya	8%	-	-	7%	15%	85%
홍시 Persimmon	19%	-	-	7%	15%	85%
배 Pear	9%	-	-	6%	15%	85%
블루베리 Blueberry	10%	-	-	7%	17%	83%
코코넛 Coconut	15%	33%	-	4%	52%	48%
두리안 Durian	27%	6%	-	7%	40%	60%
토마토 Tomato	4%	-	-	4%	8%	92%
아보카도 Avocado	1%	15%	-	10%	26%	74%

위 표는 소르베또에 주로 사용되는 과일의 평균 고형분 수치이며, 더 정확한 수치를 확인하기 위해서는 반드시 당도계에 실제 사용하는 과일의 즙을 한 방울 올려 당값을 확인해야 한다.

과일의 평균 당값으로 레시피를 작성할 경우 제조기에 넣기 전 당도계로 완성된 소르베또 믹스의 최종 당값을 확인하여 내가 설정한 최종 당값과 같은지를 확인해야 한다. 확인을 하지 않을 경우 원했던 단맛보다 더 달아지거나, 덜 달아질 수 있다.

당도계로 확인하였는데 설정한 최종 당값과 다를 경우 물이나 설탕을 추가해 조절한다. 당도계로 실제 사용하는 과일의 Brix값을 확인했다면 계량 실수가 없는 이상 설정한 최종 당값과 항상 일치한다. 하지만 그날그날 레시피를 수정해야 하는 번거로움이 있어 젤라띠에레 성향에 따라 두 가지 방법 중 선택해 소르베또를 만든다.

✦ 과일 퓌레 사용 시 제품에 표기된 Brix 값으로 과일의 당을 계산한다.

The table above shows average solids values for common fruits used in sorbetto. The check for a more accurate value, you must check the sugar value by putting a drop of juice from the fruit you will use on the refractometer.

When writing a recipe with the average sugar value of the fruit, you should check the final sugar value of the finished sorbetto mixture with a refractometer before putting it in the machine to check if it's the same as the final sugar value you set. If it's not checked, it may be sweeter or less sweet than you want it to be.

If the sugar value is different from the final value set when checked, you should adjust it by adding water or sugar. The Brix value should always match the final sugar value set if you check the Brix value of the fruit you will use with a refractometer unless there is a measurement mistake. However, there is the hassle of modifying the recipe every day, so you can choose one of two methods to make sorbetto according to gelatiere's tendency.

✦ When using fruit purée, calculate the sugar content of the fruit with the Brix value indicated on the product.

소르베또 레시피 작성 시 첫 번째로 최종 고형분 함량을 설정한다. 소르베또는 젤라또와 다르게 유제품이 들어가지 않으므로 지방과 무지유 고형분을 설정하지 않고 당이 메인 고형분 역할을 한다.

When writing a sorbetto recipe, the first step is to set the final solids content. Unlike gelato, sorbetto does not contain dairy products. Therefore, sugar serves as the main solids without setting fat and non-fat milk solids.

	당 Sugar	지방 Fat	무지유 고형분 Non-fat milk solids	기타 고형분 Other solids	총 고형분 Total solids
소르베또 Sorbetto	26~32%	-	-	0.2%↑	27~36%

두 번째로 과일의 선택과 몇 %를 넣을 것인지를 결정한다.

물과 과일을 1:1로 섞어서 마신다고 가정해보자. 입에서 느껴지는 과일의 맛이 강한 맛인지, 중간 정도의 맛인지, 약한 맛인지 구분하여 첨가량을 결정한다.

- 강한 맛 : 15 ~ 30% (예: 레몬, 라임)
- 중간 맛 : 30 ~ 55% (예: 딸기, 망고, 바나나 등 과일 중 80% 정도가 중간 맛 과일에 속한다.)
- 약한 맛 : 55 ~ 75% (예: 수박, 오렌지)

Second, decide which fruit and what percentage to use.

Suppose you drink a 1:1 mixture of water and fruit. The amount of fruit added is determined by distinguishing whether the taste in the mouth is strong, medium, or weak.

- **Strong taste** : 15~30% (e.g. : lemon, lime)
- **Medium taste** : 30~55% (e.g. : About 80% of fruits such as strawberry, mango, banana, etc. belong to medium-taste fruits.)
- **Weak taste** : 55~70% (e.g. : watermelon, orange)

세 번째로 소르베또의 과일 향을 한층 더 돋궈주는 레몬즙을 몇 %를 넣을 것인지를 결정한다.

- 레몬즙 : 0 ~ 2%

Thirdly, decide what percentage of lemon juice to add to enhance the fruity flavor of sorbetto.

- **Lemon juice** : 0~2%

마지막으로 안정제를 베이스50을 사용할 것인지, 복합안정제를 사용할 것인지에 따라 첨가량을 결정한다. 과육이 많은 과일일수록 안정제는 적게 넣고, 수분이 많을수록 안정제를 더 많이 넣어 수분을 잡아주어야 한다. 사용하는 과일의 성질을 명확하게 알아야 안정제의 적정한 첨가량을 결정할 수 있다.

- 베이스50 : 2 ~ 4%
- 복합안정제 : 0.2 ~ 0.4%

Finally, the amount of fruit added depends on whether to use Base50 or a stabilizer. The more pulps of the fruit, the less stabilizer; the more watery the fruit, the more stabilizer should be added to retain the moisture. Knowing the nature of the fruit you are using makes it possible to determine the appropriate amount of stabilizer to add.

- **Base50** : 2~4%
- **Stabilizer** : 0.2~0.4%

2 Strawberry Sorbetto with Base50 Fruits

베이스50 과일로 만드는 딸기 소르베또

딸기 소르베또 Strawberry Sorbetto

재료 Ingredients		중량 (g) Weight	당 (%) Sugar	지방 (%) Fat	무지유 고형분 (%) Non-fat milk solids	기타 고형분 (%) Other solids	총 고형분 (%) Total solids
딸기 Strawberry				-	-	(7)	
레몬즙 Lemon juice			(5)	-	-	(8)	
물 Water			-	-	-	-	
설탕 Sugar			(100)	-	-	-	
베이스50 과일 Fruit Base50			(80)	-	-	(14)	
TOTAL	g						
	%						

1. 재료와 재료의 고형분 함량을 적는다.

1. Write down the ingredients and the solids contents of the ingredients.

딸기 소르베또 Strawberry Sorbetto

재료 Ingredients		중량 (g) Weight	당 (%) Sugar	지방 (%) Fat	무지유 고형분 (%) Non-fat milk solids	기타 고형분 (%) Other solids	총 고형분 (%) Total solids
딸기 Strawberry			(8)	-	-	(7)	
레몬즙 Lemon juice			(5)	-	-	(8)	
물 Water			-	-	-	-	
설탕 Sugar			(100)	-	-	-	
베이스50 과일 Fruit Base50			(80)	-	-	(14)	
TOTAL	g	1,000 **(1)**	270 **(3)**				
	%		27% **(2)**				

2. 만들고자 하는 총 합계 중량 **(1)**과 최종 당 비율 **(2)** 그에 따른 당값 **(3)**을 설정한다.

2. Set the total weight to be made **(1)**, the final sugar ratio **(2)**, and the sugar value **(3)** accordingly.

딸기 소르베또 Strawberry Sorbetto

재료 Ingredients		중량 (g) Weight	당 (%) Sugar	지방 (%) Fat	무지유 고형분 (%) Non-fat milk solids	기타 고형분 (%) Other solids	총 고형분 (%) Total solids
딸기 Strawberry		500 **(1)**	(8) 40 **(2)**	-	-	(7) 35 **(3)**	75 **(4)**
레몬즙 Lemon juice			(5)	-	-	(8)	
물 Water			-	-	-	-	
설탕 Sugar			(100)	-	-	-	
베이스50 과일 Fruit Base50			(80)	-	-	(14)	
TOTAL	g	1,000	270				
	%		27%				

3. 과일을 몇 %로 넣을 것인지 정하고 고형분을 계산한다.

딸기는 중간 맛의 강도를 가진 과일이므로 30~55% 가이드를 가진다.

이 레시피는 50%의 비율로 계산하였다.

딸기 (당 8%, 기타 고형분 7%)

중량 : 총 중량 1,000 × 딸기 50% = 500 **(1)**

당 : 중량 500 × 당 8% = 40 **(2)**

기타 고형분 : 중량 500 × 기타 고형분 7% = 35 **(3)**

총 고형분 : 당 40 + 기타 고형분 35 = 75 **(4)**

3. Decide what percentage of fruit to put in and calculate the solids content.

Strawberry has medium-taste intensity, so that they will follow the 30~55% guide.

The recipe was calculated at a ratio of 50%.

Strawberry (8% sugar, 7% other solids)

Weight : Total weight 1,000 × Strawberry 50% = 500 **(1)**

Sugar : Weight 500 × Sugar 8% = 40 **(2)**

Other solids : Weight 500 × Other solids 7% = 35 **(3)**

Total solids : Sugar 40 + Other solids 35 = 75 **(4)**

딸기 소르베또 Strawberry Sorbetto

재료 Ingredients		중량 (g) Weight	당 (%) Sugar	지방 (%) Fat	무지유 고형분 (%) Non-fat milk solids	기타 고형분 (%) Other solids	총 고형분 (%) Total solids
딸기 Strawberry		500	(8) 40	-	-	(7) 35	75
레몬즙 Lemon juice		10 **(1)**	(5) 0.5 **(2)**	-	-	(8) 0.8 **(3)**	1.3 **(4)**
물 Water			-	-	-	-	-
설탕 Sugar			(100)	-	-	-	
베이스50 과일 Fruit Base50			(80)	-	-	(14)	
TOTAL	g	1,000	270				
	%		27%				

4. 레몬즙을 몇 %로 넣을 것인지 정하고 고형분을 계산한다.

모든 소르베또에는 레몬즙이 0 ~ 2%로 들어간다.

이 레시피는 1%의 비율로 계산하였다.

레몬 (당 5%, 기타 고형분 8%)

중량 : 총 중량 1,000 × 레몬 1% = 10 **(1)**

당 : 중량 10 × 당 5% = 0.5 **(2)**

기타 고형분 : 중량 10 × 기타 고형분 8% = 0.8 **(3)**

총 고형분 : 당 0.5 + 기타 고형분 0.8 = 1.3 **(4)**

4. Decide what percentage of lemon juice to put in and calculate the solids content.

All sorbetto contains 0~2% lemon juice.

The recipe was calculated at a ratio of 1%.

Lemon (5% sugar, 8% other solids)

Weight : Total weight 1,000 × Lemon 1% = 10 **(1)**

Sugar : Weight 10 × Sugar 5% = 0.5 **(2)**

Other solids : Weight 10 × Other solids 8% = 0.8 **(3)**

Total solids : Sugar 0.5 + Other solids 0.8 = 1.3 **(4)**

딸기 소르베또 Strawberry Sorbetto

재료 Ingredients		중량 (g) Weight	당 (%) Sugar	지방 (%) Fat	무지유 고형분 (%) Non-fat milk solids	기타 고형분 (%) Other solids	총 고형분 (%) Total solids
딸기 Strawberry		500	(8) 40	-	-	(7) 35	75
레몬즙 Lemon juice		10	(5) 0.5	-	-	(8) 0.8	1.3
물 Water			-	-	-	-	-
설탕 Sugar			(100)	-	-	-	
베이스50 과일 Fruit Base50		30 **(1)**	(80) 24 **(2)**	-	-	(14) 4.2 **(3)**	28.2 **(4)**
TOTAL	g	1,000	270				
	%		27%				

5. 과일의 성질의 파악하여 베이스50 과일의 첨가량을 결정한다.

딸기의 경우 과육과 과즙이 중간 정도로 있는 과일이므로 3%가 적정양이다.

- 질감은 개개인의 선호도가 다르므로 테스트 후에 조절한다.

베이스50 과일 (당 80%, 기타 고형분 14%)

중량 : 총 중량 1,000 × 베이스50 과일 3% = 30 **(1)**

당 : 중량 30 × 당 80% = 24 **(2)**

기타 고형분 : 중량 30 × 기타 고형분 14% = 4.2 **(3)**

총 고형분 : 당 24 + 기타 고형분 4.2 = 28.2 **(4)**

5. Determine the amount of Fruit Base50 by identifying the nature of the fruit.

3% is an appropriate amount to use for strawberries because it's a fruit with medium pulp and juice.

- Adjust the texture after tasting, as each person's preference is different.

Fruit Base50 (80% sugar, 14% other solids)

Weight : Total weight 1,000 × Fruit Base50 3% = 30 **(1)**

Sugar : Weight 30 × Sugar 80% = 24 **(2)**

Other solids : Weight 30 × Other solids 14% = 4.2 **(3)**

Total solids : Sugar 24 + Other solids 4.2 = 28.2 **(4)**

딸기 소르베또 Strawberry Sorbetto

재료 Ingredients		중량 (g) Weight	당 (%) Sugar	지방 (%) Fat	무지유 고형분 (%) Non-fat milk solids	기타 고형분 (%) Other solids	총 고형분 (%) Total solids
딸기 Strawberry		500	(8) 40	-	-	(7) 35	75
레몬즙 Lemon juice		10	(5) 0.5	-	-	(8) 0.8	1.3
물 Water			-	-	-	-	-
설탕 Sugar		206 **(2)**	(100) 205.5 **(1)**	-	-	-	205.5 **(3)**
베이스50 과일 Fruit Base50		30	(80) 24	-	-	(14) 4.2	28.2
TOTAL	g	1,000	270				
	%		27%				

6. 당의 총 무게에서 딸기, 레몬즙, 베이스50 과일의 당을 뺀 값이 설탕의 당량이 된다.

설탕 (당 100%)

당 : 최종 당 270 – 딸기의 당 40 – 레몬즙의 당 0.5 – 베이스50 과일의 당 24 = 205.5

설탕은 100% 고형분이므로 당량 **(1)**, 중량 **(2)**, 총 고형분량 **(3)** 이 동일하다.

6. The value obtained by subtracting the sugar of strawberry, lemon juice, and Fruit Base50 from the total weight of sugar is sugar weight.

Sugar (100% sugar)

Sugar : Final sugar 270 - Strawberry's sugar 40 - Lemon juice's sugar 0.5 - Fruit Base50's sugar 24 = 205.5

Since sugar is 100% solids, sugar weight **(1)**, weight **(2)**, and total solids **(3)** are the same.

딸기 소르베또 Strawberry Sorbetto

재료 Ingredients		중량 (g) Weight	당 (%) Sugar	지방 (%) Fat	무지유 고형분 (%) Non-fat milk solids	기타 고형분 (%) Other solids	총 고형분 (%) Total solids
딸기 Strawberry		500	(8) 40	-	-	(7) 35	75
레몬즙 Lemon juice		10	(5) 0.5	-	-	(8) 0.8	1.3
물 Water		254 **(1)**	-	-	-	-	-
설탕 Sugar		206	(100) 205.5	-	-	-	205.5
베이스50 과일 Fruit Base50		30	(80) 24	-	-	(14) 4.2	28.2
TOTAL	g	1,000	270				
	%		27%				

7. 마지막 물은 총 중량에서 물을 제외한 모든 재료의 중량을 뺀 값이다.

물 (수분 100%)

중량 : 총 중량 1,000 – 딸기 500 – 레몬즙 10 – 설탕 206 – 베이스50 과일 30 = 254 **(1)**

7. Water is the total weight minus the weight of ingredients except water.

Water (100% water)

Weight : Total weight 1,000 - Strawberry 500 - Lemon juice 10 - Sugar 206 - Fruit Base50 30 = 254 **(1)**

딸기 소르베또 Strawberry Sorbetto

재료 Ingredients		중량 (g) Weight	당 (%) Sugar	지방 (%) Fat	무지유 고형분 (%) Non-fat milk solids	기타 고형분 (%) Other solids	총 고형분 (%) Total solids
딸기 Strawberry		500	(8) 40	-	-	(7) 35	75
레몬즙 Lemon juice		10	(5) 0.5	-	-	(8) 0.8	1.3
물 Water		254	-	-	-	-	-
설탕 Sugar		206	(100) 205.5	-	-	-	205.5
베이스50 과일 Fruit Base50		30	(80) 24	-	-	(14) 4.2	28.2
TOTAL	g	1,000	270			40 **(1)**	310 **(1)**
	%		27%			4% **(2)**	31% **(2)**

8. 기타 고형분과 총 고형분값도 모두 더해 비율을 계산한다.

기타 고형분 : 딸기 35 + 레몬즙 0.8 + 베이스50 과일 4.2 = 40 **(1)**

기타 고형분 (%) : 40 ÷ 1,000% = 4 **(2)**

총 고형분 : 딸기 75 + 레몬즙 1.3 + 설탕 205.5 + 베이스50 과일 28.2 = 310 **(1)**

총 고형분 (%) : 310 ÷ 1,000% = 31 **(2)**

8. Calculate the ratio by adding all other solids and total solids values.

Other solids : Strawberry 35 + Lemon juice 0.8 + Fruit Base50 4.2 = 40 **(1)**

Other solids (%) : 40 ÷ 1,000% = 4 **(2)**

Total solids : Strawberry 75 + Lemon juice 1.3 + Sugar 205.5 + Fruit Base50 28.2 = 310 **(1)**

Total solids (%) : 310 ÷ 1,000% = 31 **(2)**

3 Strawberry Sorbetto made with Stabilizer

복합안정제로 만드는 딸기 소르베또

딸기 소르베또 Strawberry Sorbetto

재료 Ingredients		중량 (g) Weight	당 (%) Sugar	지방 (%) Fat	무지유 고형분 (%) Non-fat milk solids	기타 고형분 (%) Other solids	총 고형분 (%) Total solids
딸기 Strawberry			(8)	-	-	(7)	
레몬즙 Lemon juice			(5)	-	-	(8)	
물 Water			-	-	-	-	
설탕 Sugar			(100)	-	-	-	
포도당 Dextrose			(92)	-	-	-	
복합안정제 Stabilizer			-	-	-	(100)	
TOTAL	g						
	%						

1. 재료와 재료의 고형분 함량을 적는다.

1. Write down the ingredients and the solids contents of the ingredients.

딸기 소르베또 Strawberry Sorbetto

재료 Ingredients		중량 (g) Weight	당 (%) Sugar	지방 (%) Fat	무지유 고형분 (%) Non-fat milk solids	기타 고형분 (%) Other solids	총 고형분 (%) Total solids
딸기 Strawberry			(8)	-	-	(7)	
레몬즙 Lemon juice			(5)	-	-	(8)	
물 Water			-	-	-	-	
설탕 Sugar			(100)	-	-	-	
포도당 Dextrose			(92)	-	-	-	
복합안정제 Stabilizer			-	-	-	(100)	
TOTAL	g	1,000 **(1)**	270 **(3)**				
	%		27% **(2)**				

2. 만들고자 하는 총 합계 중량 **(1)**과 최종 당 비율 **(2)**과 그에 따른 당값 **(3)**을 설정한다.

2. Set the total weight to be made **(1)**, the final sugar ratio **(2)**, and the sugar value **(3)** accordingly.

딸기 소르베또 Strawberry Sorbetto

재료 Ingredients		중량 (g) Weight	당 (%) Sugar	지방 (%) Fat	무지유 고형분 (%) Non-fat milk solids	기타 고형분 (%) Other solids	총 고형분 (%) Total solids
딸기 Strawberry		500 **(1)**	(8) 40 **(2)**	-	-	(7) 35 **(3)**	75 **(4)**
레몬즙 Lemon juice			(5)	-	-	(8)	
물 Water			-	-	-	-	-
설탕 Sugar			(100)	-	-	-	
포도당 Dextrose			(92)	-	-	-	
복합안정제 Stabilizer			-	-	-	(100)	
TOTAL	g	1,000	270				
	%		27%				

3. 과일을 몇 %로 넣을 것인지 정하고 고형분을 계산한다.

딸기는 중간 맛의 강도를 가진 과일이므로 30~55% 가이드를 가진다.

이 레시피는 50%의 비율로 계산하였다.

딸기 (당 8%, 기타 고형분 7%)

중량 : 총 중량 1,000 × 딸기 50% = 500 **(1)**

당 : 중량 500 × 당 8% = 40 **(2)**

기타 고형분 : 중량 500 × 기타 고형분 7% = 35 **(3)**

총 고형분 : 당 40 + 기타 고형분 35 = 75 **(4)**

3. Decide what percentage of fruit to put in and calculate the solids content.

Strawberry has medium-taste intensity, so they will follow the 30~55% guide.

The recipe was calculated at a ratio of 50%.

Strawberry (8% sugar, 7% other solids)

Weight : Total weight 1,000 × Strawberry 50% = 500 **(1)**

Sugar : Weight 500 × Sugar 8% = 40 **(2)**

Other solids : Weight 500 × Other solids 7% = 35 **(3)**

Total solids : Sugar 40 + Other solids 35 = 75 **(4)**

딸기 소르베또 Strawberry Sorbetto

재료 Ingredients		중량 Weight	당 Sugar	지방 Fat	무지유 고형분 Non-fat milk solids	기타 고형분 Other solids	총 고형분 Total solids
딸기 Strawberry		500	(8) 40	-	-	(7) 35	75
레몬즙 Lemon juice		10 **(1)**	(5) 0.5 **(2)**	-	-	(8) 0.8 **(3)**	1.3 **(4)**
물 Water			-	-	-	-	-
설탕 Sugar			(100)	-	-	-	
포도당 Dextrose			(92)	-	-	-	
복합안정제 Stabilizer			-	-	-	(100)	
TOTAL	g	1,000	270				
	%		27%				

4. 레몬즙을 몇 %로 넣을 것인지 정하고 고형분을 계산한다.

모든 소르베또는 레몬즙이 0 ~ 2% 들어간다.

이 레시피는 1%의 비율로 계산하였다.

레몬 (당 5%, 기타 고형분 8%)

중량 : 총 중량 1,000 × 레몬 1% = 10 **(1)**

당 : 중량 10 × 당 5% = 0.5 **(2)**

기타 고형분 : 중량 10 × 기타 고형분 8% = 0.8 **(3)**

총 고형분 : 당 0.5 + 기타 고형분 0.8 = 1.3 **(4)**

4. Decide what percentage of Lemon juice to put in and calculate the solids content.

All sorbetto contains 0~2% lemon juice.

The recipe was calculated at a ratio of 1%.

Lemon (5% sugar, 8% other solids)

Weight : Total weight 1,000 × Lemon 1% = 10 **(1)**

Sugar : Weight 10 × Sugar 5% = 0.5 **(2)**

Other solids : Weight 10 × Other solids 8% = 0.8 **(3)**

Total solids : Sugar 0.5 + Other solids 0.8 = 1.3 **(4)**

딸기 소르베또 Strawberry Sorbetto

재료 Ingredients		중량 (g) Weight	당 (%) Sugar	지방 (%) Fat	무지유 고형분 (%) Non-fat milk solids	기타 고형분 (%) Other solids	총 고형분 (%) Total solids
딸기 Strawberry		500	(8) 40	-	-	(7) 35	75
레몬즙 Lemon juice		10	(5) 0.5	-	-	(8) 0.8	1.3
물 Water			-	-	-	-	-
설탕 Sugar			(100)	-	-	-	
포도당 Dextrose			(92)	-	-	-	
복합안정제 Stabilizer		3 **(1)**	-	-	-	(100) 3 **(2)**	3 **(3)**
TOTAL	g	1,000	270				
	%		27%				

5. 과일의 성질의 파악하여 복합안정제의 첨가량을 결정한다.

딸기의 경우 과육과 과즙이 중간 정도로 있는 과일이므로 0.3%가 적정양이다.

- 질감은 개개인의 선호도가 다르므로 테스트 후 조절한다.

복합안정제 (기타 고형분 100%)

중량 : 총 중량 1,000 × 복합안정제 0.3% = 3 **(1)**

복합안정제는 100% 고형분이므로 기타 고형분 **(2)**, 총 고형분 **(3)**이 중량과 동일하다.

5. Determine the amount of Fruit Base50 by identifying the nature of the fruit.

0.3% is an appropriate amount to use for strawberries because it's a fruit with medium pulp and juice.

- Adjust the texture after tasting, as each person's preference is different.

Stabilizer (100% other solids)

Weight : Total weight 1,000 × stabilizer 0.3%
= 3 **(1)**

Since stabilizer is 100% solids, weight of other solids **(2)**, and total solids **(3)** are the same.

딸기 소르베또 Strawberry Sorbetto

재료 Ingredients		중량 (g) Weight	당 (%) Sugar	지방 (%) Fat	무지유 고형분 (%) Non-fat milk solids	기타 고형분 (%) Other solids	총 고형분 (%) Total solids
딸기 Strawberry		500	(8) 40	-	-	(7) 35	75
레몬즙 Lemon juice		10	(5) 0.5	-	-	(8) 0.8	1.3
물 Water			-	-	-	-	-
설탕 Sugar		211 **(2)**	(100) 211.14 **(1)**	-	-	-	211.14 **(3)**
포도당 Dextrose		20 **(2)**	(92) 18.36 **(1)**	-	-	-	18.36 **(3)**
복합안정제 Stabilizer		3	-	-	-	(100) 3	3
TOTAL	g	1,000	270				
	%		27%				

6. 당의 총 무게에서 딸기와 레몬즙의 당을 뺀 값이 설탕과 포도당의 당량이 된다.

소르베또에서 설탕과 포도당의 당 비율을 나눌 때 설탕은 최소 80%, 포도당은 최대 20% 안에서 나눠 준다.

이 레시피는 설탕 92%, 포도당 8% 비율의 예시이다.

당의 총 무게 270에서 딸기와 레몬즙의 당 40.5를 뺀 229.5에서 비율을 계산한다.

설탕 (당 100%)

당 : 229.5 × 92% = 211.14

→ 설탕은 100% 고형분이므로 당량 **(1)**, 중량 **(2)**, 총 고형분량 **(3)**이 동일하다.

포도당 (당 92%, 수분 8%)

당 : 229.5 × 8% = 18.36 **(1)**

→ 포도당은 8% 수분을 포함하므로 중량은 8% 수분이 포함된 값을 계산한다.

중량 : 18.36 ÷ 당 92% = 19.9 **(2)**

총 고형분 : 18.36 **(3)**

6. The value obtained by subtracting the sugar of strawberry and lemon juice from the total weight of sugar is sugar and dextrose weight.

When dividing the sugar ratio of sugar and dextrose in sorbetto, sugar is at least 80%, and dextrose is divided within 20%.

This recipe is an example of a 92% sugar to 8% dextrose ratio.

Calculate the ratio from 229.5 by subtracting 40.5 of the sugar in the strawberry and lemon juice from the total weight of sugar at 270.

Sugar (100% sugar)

Sugar : 229.5 × 92% = 211.14

→ Since sugar is 100% solids, sugar weight **(1)**, weight **(2)**, and total solids **(3)** are the same.

Dextrose (92% sugar, 8% water)

Sugar : 229.5 × 8% = 18.36 **(1)**

→ Since dextrose contains 8% water, the weight is calculated as containing 8% water.

Weight : 18.36 ÷ Sugar 92% = 19.9 **(2)**

Total solids : 18.36 **(3)**

딸기 소르베또 Strawberry Sorbetto

재료 Ingredients		중량 (g) Weight	당 (%) Sugar	지방 (%) Fat	무지유 고형분 (%) Non-fat milk solids	기타 고형분 (%) Other solids	총 고형분 (%) Total solids
딸기 Strawberry		500	(8) 40	-	-	(7) 35	75
레몬즙 Lemon juice		10	(5) 0.5	-	-	(8) 0.8	1.3
물 Water		256 **(1)**	-	-	-	-	-
설탕 Sugar		211	(100) 211.14	-	-	-	211.14
포도당 Dextrose		20	(92) 18.36	-	-	-	18.36
복합안정제 Stabilizer		3	-	-	-	(100) 3	3
TOTAL	g	1,000	270				
	%		27%				

7. 마지막 물은 총 중량에서 물을 제외한 모든 재료의 중량을 뺀 값이다.

물 (수분 100%)

중량 : 총 중량 1,000 – 딸기 500 - 레몬즙 10 – 설탕 211 – 포도당 20 – 복합안정제 3 = 256 **(1)**

7. The final water is the total weight minus the weight of all the ingredients except the water.

Water (100% water)

Weight : Total weight 1,000 - Strawberry 500 - Lemon juice 10 - Sugar 211 - Dextrose 20 - Stabilizer 3 = 256 **(1)**

딸기 소르베또 Strawberry Sorbetto

재료 Ingredients		중량 (g) Weight	당 (%) Sugar	지방 (%) Fat	무지유 고형분 (%) Non-fat milk solids	기타 고형분 (%) Other solids	총 고형분 (%) Total solids
딸기 Strawberry		500	(8) 40	-	-	(7) 35	75
레몬즙 Lemon juice		10	(5) 0.5	-	-	(8) 0.8	1.3
물 Water		256	-	-	-	-	-
설탕 Sugar		211	(100) 211.14	-	-	-	211.14
포도당 Dextrose		20	(92) 18.36	-	-	-	18.36
복합안정제 Stabilizer		3	-	-	-	(100) 3	3
TOTAL	g	1,000	270			38.8 **(1)**	308.8 **(1)**
	%		27%			3.9% **(2)**	30.9% **(2)**

8. 기타 고형분과 총 고형분값도 모두 더해 비율을 계산한다.

기타 고형부 : 딸기 35 + 레몬즙 0.8 + 복합안정제 3 = 38.8 **(1)**

기타 고형분 (%) : 38.8 ÷ 1,000% = 3.9 **(2)**

총 고형분 : 딸기 75 + 레몬즙 1.3 + 설탕 211.14 + 포도당 18.36 + 복합안정제 3 = 308.8 **(1)**

총 고형분 (%) : 308.8 ÷ 1,000% = 30.9 **(2)**

8. Calculate the ratio by adding all other solids and total solids values.

Other solids : Strawberry 35 + Lemon juice 0.8 + Stabilizer 3 = 38.8 **(1)**

Other solids (%) : 38.8 ÷ 1,000% = 3.9 **(2)**

Total solids : Strawberry 75 + Lemon juice 1.3 + Sugar 211.14 + Dextrose 18.36 + Stabilizer 3 = 308.8 **(1)**

Total solids (%) : 308.8 ÷ 1,000% = 30.9 **(2)**

4 Lemon Sorbetto made with Fruit Base50

베이스50 과일로 만드는 레몬 소르베또

레몬 소르베또 Lemon Sorbetto

재료 Ingredients		중량 (g) Weight	당 (%) Sugar	지방 (%) Fat	무지유 고형분 (%) Non-fat milk solids	기타 고형분 (%) Other solids	총 고형분 (%) Total solids
레몬 Lemon		150	(5) 7.5	-	-	(8) 12	19.5
레몬즙 Lemon juice		-	-	-	-	-	-
물 Water		549	-	-	-	-	-
설탕 Sugar		261	(100) 260.5	-	-	-	260.5
베이스50 과일 Fruit Base50		40	(80) 32	-	-	(14) 5.6	37.6
TOTAL	g	1,000	300			17.6	317.6
	%		30%			1.8%	31.8%

✦ 레몬은 강한 맛의 강도를 가지므로 레몬의 첨가량은 15 ~ 30%이다.

✦ Because the lemon has a strong-taste intensity, 15~30% will be added.

5

Lemon Sorbetto made with Stabilizer

복합안정제로 만드는 레몬 소르베또

레몬 소르베또 Lemon Sorbetto

재료 Ingredients		중량 (g) Weight	당 (%) Sugar	지방 (%) Fat	무지유 고형분 (%) Non-fat milk solids	기타 고형분 (%) Other solids	총 고형분 (%) Total solids
레몬 Lemon		150	(5) 7.5	-	-	(8) 12	19.5
레몬즙 Lemon juice		-	-	-	-	-	-
물 Water		548	-	-	-	-	
설탕 (80%) Sugar (80%)		234	(100) 234	-	-	-	234
포도당 (20%) Dextrose (20%)		64	(92) 58.5	-	-	-	58.5
복합안정제 Stabilizer		4	-	-	-	(100) 4	4
TOTAL	g	1,000	300			16	316
	%		30%			1.6%	31.6%

6 Watermelon Sorbetto made with Fruit Base50

베이스50 과일로 만드는 수박 소르베또

수박 소르베또 Watermelon Sorbetto

재료 Ingredients		중량 Weight	당 (%) Sugar	지방 (%) Fat	무지유 고형분 (%) Non-fat milk solids	기타 고형분 (%) Other solids	총 고형분 (%) Total solids
수박 Watermelon		700	(6) 42	-	-	(5) 3.5	77
레몬즙 Lemon juice		10	(5) 0.5	-	-	(8) 0.8	1.3
물 Water		28	-	-	-	-	-
설탕 Sugar		227	(100) 229.5	-	-	-	229.5
베이스50 과일 Fruit Base50		35	(80) 28	-	-	(14) 4.9	32.9
TOTAL	g	1,000	300			40.7	340.7
	%		30%			4%	34.1%

✦ 수박은 약한 맛의 강도를 가지므로 첨가량은 55 ~ 75%이다.

✦ Because the watermelon has a wear-taste intensity, 55~75% will be added.

7

Watermelon Sorbetto made with Stabilizer

복합안정제로 만드는 수박 소르베또

수박 소르베또 Watermelon Sorbetto

재료 Ingredients		중량 Weight	당 (%) Sugar	지방 (%) Fat	무지유 고형분 (%) Non-fat milk solids	기타 고형분 (%) Other solids	총 고형분 (%) Total solids
수박 Watermelon		700	(6) 42	-	-	(5) 35	77
레몬즙 Lemon juice		10	(5) 0.5	-	-	(8) 0.8	1.3
물 Water		24	-	-	-	-	
설탕 (80%) Sugar (80%)		206	(100) 206	-	-	-	206
포도당 (20%) Dextrose (20%)		56	(92) 51.5	-	-	-	51.5
복합안정제 Stabilizer		3.5	-	-	-	(100) 3.5	3.5
TOTAL	g	1,000	300			39.3	339.3
	%		30%			3.9%	33.9%

✦ 약한 맛의 강도를 가진 과일은 과일 첨가량에 따라 물은 한방울도 들어가지 않고 과일, 당, 안정제만으로도 소르베또를 만들 수 있다.

✦ 레몬이나 수박 등 수분이 많은 과일은 이눌린을 1~4% 정도 첨가하면 보다 더 부드러운 질감으로 조절할 수 있다.

✦ For the fruits with weak taste intensity, you can make sorbetto with only fruit, sugar, and stabilizers without adding a drop of water, depending on the amount of fruit added.

✦ For the fruits with a lot of water, such as lemons and watermelons, you can adjust to a softer texture by adding 1~4% of inulin.

Sorbetto POD & PAC

(Potere Dolcificante) (Poter Anti Congelante)

소르베또 POD(감미도) & PAC(빙점강하력)

• POD와 PAC를 계산하기 위해서는 47~51p 표의 수치를 참고한다.
• To calculate POD and PAC, refer to the table on pages 47~51.

소르베또 POD Sorbetto POD (Potere Dolcificante/ SP: Sweetening Power)

딸기 소르베또 Strawberry Sorbetto

재료 Ingredients		중량 (g) Weight	당 (%) Sugar	지방 (%) Fat	무지유 고형분 (%) Non-fat milk solids	기타 고형분 (%) Other solids	총 고형분 (%) Total solids	POD (SP)	PAC (AFP)
딸기 Strawberry		500	(8) 40	-	-	(7) 35	75	52 **(1)**	
레몬즙 Lemon juice		10	(5) 0.5	-	-	(8) 0.8	1.3	0.65 **(2)**	
물 Water		256	-	-	-	-	-	-	-
설탕 Sugar		211	(100) 211.14	-	-	-	211.14	211.14 **(3)**	-
포도당 Dextrose		20	(92) 18.36	-	-	-	18.36	13.2 **(4)**	-
복합안정세 Stabilizer		3	-	-	-	(100) 3	3	-	-
TOTAL	g	1,000	270			38.8	308.8	276 **(5)**	
	%		27%			3.9%	30.9%	27.6 **(6)**	

1. **딸기**
   POD = 딸기 당 40 × 과당 POD 1.3 = 52 **(1)**
2. **레몬즙**
   POD = 레몬즙 당 0.5 × 과당 POD 1.3 = 0.65 **(2)**
3. **설탕**
   POD = 설탕 당량 211.14 × 설탕 POD 1 = 211.14 **(3)**
4. **함수결정포도당**
   POD = 포도당 당량 18.36 × 포도당 POD 0.72 = 13.2 **(4)**
5. 모든 POD 값을 더한다.
   POD = 딸기 52 + 레몬즙 0.65 + 설탕 211.14 + 포도당 13.2
   = 276 **(5)**
6. POD 276 ÷ 1,000%= 27.6 **(6)**
   젤라또와 다르게 소르베또에서 최종 당의 %보다 POD값이 높을 수 있는 이유는 과일 안에 함유된 과당의 영향 때문이다. 소르베또는 어떤 과일을 몇 %로 사용하는지에 따라 POD와 PAC 수치 도출 값이 다양하게 나타날 수 있다.

1. **Strawberry**
   POD = Sugar of strawberry 40 × Fructose POD 1.3 = 52 **(1)**
2. **Lemon juice**
   POD = Sugar of lemon juice 0.5 × Fructose POD 1.3
   = 0.65 **(2)**
3. **Sugar**
   POD = Sugar weight of sugar 211.14 × Sugar POD 1
   = 211.14 **(3)**
4. **Dextrose**
   POD = Sugar weight of dextrose 18.36 ×
   Dextrose POD 0.72 = 13.2 **(4)**
5. Add all the POD values.
   POD = Strawberry 52 + Lemon juice 0.65 + Sugar 211.14 +
   Dextrose 13.2 = 276 **(5)**
6. POD 276 ÷ 1,000% = 27.6 **(6)**
   Unlike gelato, the reason why the POD value can be higher than the percentage of final sugar in sorbetto is due to the effect of fructose contained in the fruit. The derived POD and PAC values may vary depending on what kind of fruit is used in sorbetto and what percentage.

소르베또 PAC Sorbetto POD (Potere Dolcificante/ SP: Sweetening Power)

딸기 소르베또 Strawberry Sorbetto

재료 Ingredients		중량 (g) Weight	당 (%) Sugar	지방 (%) Fat	무지유 고형분 (%) Non-fat milk solids	기타 고형분 (%) Other solids	총 고형분 (%) Total solids	POD (SP)	PAC (AFP)
딸기 Strawberry		500	(8) 40	-	-	(7) 35	75	52	76 **(1)**
레몬즙 Lemon juice		10	(5) 0.5	-	-	(8) 0.8	1.3	0.65	0.95 **(2)**
물 Water		256	-	-	-	-	-	-	
설탕 Sugar		211	(100) 211.14	-	-	-	211.14	211.14	211.14 **(3)**
포도당 Dextrose		20	(92) 18.36	-	-	-	18.36	13.2	34.9 **(4)**
복합안정제 Stabilizer		3	-	-	-	(100) 3	3	-	-
TOTAL	g	1,000	270			38.8	308.8	276	323 **(5)**
	%		27%			3.9%	30.9%	27.6	32.3 **(6)**

1. **딸기**
   PAC = = 딸기 당 40 × 과당 PAC 1.9 = 76 **(1)**
2. **레몬즙**
   PAC = 레몬즙 당 0.5 × 과당 PAC 1.9 = 0.95 **(2)**
3. **설탕**
   PAC = 설탕 당량 211.14 × 설탕 PAC 1 = 211.14 **(3)**
4. **함수결정포도당**
   PAC = 포도당 당량 18.36 × 포도당 PAC 1.9 = 34.9 **(4)**
5. 모든 PAC값을 더한다.
   PAC = 딸기 76 + 레몬즙 0.95 + 설탕 211.14 + 포도당 34.9
   = 323 **(5)**
6. PAC 323 ÷ 1,000%= 32.3 **(6)**
   도출된 PAC값에 ÷ 2.5를 하여 -를 붙인다. 그러면 내 레시피와 어울리는 기본 쇼케이스 온도가 된다.
   이 경우 32.3 ÷ 2.5 = 12.9
   즉, -12.9°C가 이 레시피와 어울리는 기본 온도이다.

1. **Strawberry**
   PAC = Sugar of strawberry 40 × Fructose PAC 1.9 = 76 **(1)**
2. **Lemon juice**
   PAC = Sugar of lemon juice 0.5 × Fructose PAC 1.9 = 0.95 **(2)**
3. **Sugar**
   PAC = Sugar weight of sugar 211.14 × Sugar PAC 1
   = 211.14 **(3)**
4. **Dextrose**
   PAC = Sugar weight of dextrose 18.36 × Dextrose PAC 1.9
   = 34.9 **(4)**
5. Add all the PAC values.
   PAC = Strawberry 76 + Lemon juice 0.95 + Sugar 211.14 + Dextrose 34.9 = 323 **(5)**
6. PAC 323 ÷ 1,000% = 32.3 **(6)**
   Divide the derived PAC value by 2.5 and add - (negative). This will be the default showcase temperature that matches the recipe.
   In the above example, 32.3 ÷ 2.5 = 12.9
   Therefore, -12.9°C is the primary temperature for the above recipe.

	POD	PAC
젤라또 Gelato	12 - 22	23 - 30
소르베또 Sorbetto	20 - 30	28 - 36

Part 7.

Sorbetto Recipes

소르베또 레시피

1 Phytochemical Sorbetto

피토케미컬 소르베또

'피토케미컬Phytochemical'은 식물을 뜻하는 'Phyto'와 화학을 뜻하는 'Chemical'의 합성어로 채소나 과일의 색, 향, 맛, 영양 효과를 결정하고 해충이나 미생물로부터 스스로를 보호하기 위해 만들어내는 중요한 보호 물질이다. 피토케미컬을 섭취하면 항산화 효과는 물론 면역력 향상 등 건강에 좋은 영향을 미친다. 여기에서는 1만여 종이 넘는 피토케미컬 종류 중 소르베또에 어울리는 안토시아닌(트리플베리), 리코펜(딸기), 베타카로틴(오렌지), 엽록소(키위), 플라보노이드(배)가 함유된 과일을 사용했다.

'Phytochemical' is a compound word of 'phyto' meaning plant, and 'chemical' for artificially prepared substance. It is an essential protective substance that determines the color, aroma, taste, and nutritional effect of vegetables or fruits and is created to protect itself from pests or microorganisms. Intake of phytochemicals has sound health effects, such as antioxidant effect and immunity improvement. Among more than 10,000 types of phytochemicals, fruits containing anthocyanin (triple berry), lycopene (strawberry), beta-carotene (orange), chlorophyll (kiwi), and flavonoid (pear) suitable for sorbetto were used.

Ingredients		Quantity
트리플베리	Triple Berry	150g
키위	Kiwi	120g
오렌지즙	Orange juice	247g
배	Pear	170g
딸기	Strawberries	150g
설탕	Sugar	120g
함수결정포도당	Dextrose	30g
이눌린	Inulin	10g
복합안정제	Stabilizer	3g
TOTAL		1,000g

1. 비커에 트리플베리, 키위, 오렌지즙, 배, 딸기를 계량한다.
2. 믹싱볼에 설탕, 함수결정포도당, 이눌린, 복합안정제를 계량한 후 스패출러로 혼합한다.
3. 2를 1에 넣고 핸드블렌더로 믹싱한다.
4. 제조기에 넣고 냉각 교반한다.

1. Measure triple berries, kiwi, orange juice, pear, and strawberries in a container.
2. Weight all the powder ingredients- sugar, dextrose, inulin, and stabilizer in a mixing bowl and combine with a spatula.
3. Mix 2 with 1 and mix using a hand blender.
4. Pour into the machine and cold-churn the mixture.

당 Sugar	지방 Fat	무지유 고형분 Non-fat milk solids	기타 고형분 Other solids	총 고형분 Total solids	수분 Water
22.8%	-	-	4.9%	27.7%	72.3%

2 Detox Sorbetto

디톡스 소르베또

체내에 축적된 독소를 제거하고 건강한 상태로 되돌리는 의미의 '디톡스Detox'는 주스나 차 등 다양한 방법으로 접할 수 있다. 완전한 디톡스라고는 할 수 없지만 디톡스의 개념을 소르베또에 접목시켜 바나나, 파인애플, 오이, 샐러리, 생강, 레몬을 혼합해 만들어보았다.

Detox, which means removing toxins accumulated in the body and returning to a healthy state, can be accessed in various forms, such as juice or tea. It may not be a complete detox product, but I grafted the concept of detox into sorbetto and made it by mixing banana, pineapple, cucumber, celery, ginger, and lemon.

Ingredients		Quantity
오이즙	Cucumber juice	153g
샐러리즙	Celery juice	80g
생강즙	Ginger juice	15g
레몬즙	Lemon juice	10g
바나나	Banana	305g
파인애플	Pineapple	285g
설탕	Sugar	120g
함수결정포도당	Dextrose	30g
복합안정제	Stabilizer	2g
TOTAL		1,000g

1. 착즙기를 사용해 오이, 샐러리, 생강, 레몬을 착즙한 후 바나나, 파인애플과 함께 비커에 계량한다.
2. 믹싱볼에 설탕, 함수결정포도당, 복합안정제를 계량한 후 스패출러로 혼합한다.
3. 2를 1에 넣고 핸드블렌더로 믹싱한다.
4. 제조기에 넣고 냉각 교반한다.

1. Using a juicer, extract cucumber, celery, ginger, and lemon and measure them with banana and pineapple in a container.
2. Weight sugar, dextrose, and stabilizer in a mixing bowl and combine with a spatula.
3. Mix 2 with 1 and mix using a hand blender.
4. Pour into the machine and cold-churn the mixture.

당 Sugar	지방 Fat	무지유 고형분 Non-fat milk solids	기타 고형분 Other solids	총 고형분 Total solids	수분 Water
22.8%	-	-	6.4%	29.1%	70.9%

3 Citrus Sinfonia Sorbetto

시트러스 신포니아 소르베토

한 가지 과일이 아닌 여러 가지 과일을 혼합해 사용할 때는 어떤 과일의 맛을 강조할 것인지, 또는 사용하는 모든 과일이 조화롭게 어우러지게 할 것인지를 잘 판단해야 한다. 여기에서는 한 가지 과일을 강조하기보다는 시트러스 계열의 다양한 과일의 조화로운 하모니를 느낄 수 있게 하였다.

When using a mixture of more than one fruit, you must decide carefully which fruit to emphasize or whether to harmonize all the fruits you use. Here, rather than emphasizing one fruit, I made it to feel the harmonious balance of various citrus fruits.

Ingredients		Quantity
레몬 퓌레	Lemon purée	20g
라임 퓌레	Lime purée	30g
자몽 퓌레	Grapefruit purée	300g
만다린 퓌레	Mandarin purée	250g
베르가못 퓌레	Bergamot purée	150g
설탕	Sugar	180g
함수결정포도당	Dextrose	46g
이눌린	Inulin	20g
복합안정제	Stabilizer	4g
TOTAL		1,000g

1. 모든 과일 퓌레는 냉장고에 두어 해동시킨 후 비커에 계량한다.
2. 믹싱볼에 설탕, 함수결정포도당, 이눌린, 복합안정제를 계량한 후 스패출러로 혼합한다.
3. 2를 1에 넣고 핸드블렌더로 믹싱한다.
4. 제조기에 넣고 냉각 교반한다.

● 과일 퓌레 대신 생과일을 사용할 경우 착즙기를 사용하여 착즙한다.

1. Defrost all the purée in the chiller overnight. Measure them with lemon juice and water in a container.
2. Weight sugar, dextrose, and stabilizer in a mixing bowl and combine with a spatula.
3. Mix 2 with 1 and mix using a hand blender.
4. Pour into the machine and cold-churn the mixture.

● When using fresh fruits instead of purée, use a juicer to extract the juice.

당	지방	무지유 고형분	기타 고형분	총 고형분	수분
Sugar	Fat	Non-fat milk solids	Other solids	Total solids	Water
30%	-	-	2%	32%	68%

4 **Tropical Sorbetto**

트로피컬 소르베또

기후 변화로 인해 한국도 점점 아열대성 기후로 바뀌고 있어 열대과일을 재배하는 농가도 그만큼 늘어나고 있다. 국내 열대과일 재배 면적 1위인 망고와 2위인 패션푸르트를 사용해 달콤하면서도 상큼한 맛으로 포인트를 주고, 코코넛으로 녹진한 부드러움을 더해보았다.

Due to climate change, Korea is gradually changing to a subtropical climate, and the number of farmhouses growing tropical fruits is also increasing. Mango has the number one tropical fruit cultivation area in Korea, and passion fruit, which comes in second, gives an impact with a sweet yet refreshing taste, and coconut is used to add softness.

Ingredients		Quantity
망고 퓌레	Mango purée	200g
패션푸르트 퓌레	Passion fruit purée	100g
코코넛 퓌레	Coconut purée	150g
레몬즙	Lemon juice	10g
물	Water	333g
설탕	Sugar	160g
함수결정포도당	Dextrose	45g
복합안정제	Stabilizer	2g
TOTAL		1,000g

1. 모든 과일 퓌레는 냉장고에 두어 해동시킨 후 레몬즙, 물과 함께 비커에 계량한다.
2. 믹싱볼에 설탕, 함수결정포도당, 복합안정제를 계량한 후 스패출러로 혼합한다.
3. 2를 1에 넣고 핸드블렌더로 믹싱한다.
4. 제조기에 넣고 냉각 교반한다.

1. Defrost all the purée in the chiller overnight. Measure them with lemon juice and water in a container.
2. Weight sugar, dextrose, and stabilizer in a mixing bowl and combine with a spatula.
3. Mix 2 with 1 and mix using a hand blender.
4. Pour into the machine and cold-churn the mixture.

당 Sugar	지방 Fat	무지유 고형분 Non-fat milk solids	기타 고형분 Other solids	총 고형분 Total solids	수분 Water
28.6%	3.3%	-	0.6%	32.5%	67.5%

5 Peach & Black Tea Sorbetto

Hot Infusion

피치 & 홍차 소르베또

복숭아와 홍차는 디저트나 요리 분야에서 많이 사용되는 조합 중 하나로, 그만큼 많은 사람들이 거부감 없이 즐길 수 있는 조합이다. 좋아하는 종류의 홍차를 선택하거나 복숭아 생과 대신 퓌레를 사용해도 좋다.

Peach and black tea are one of the most used flavor combinations in desserts and cooking, and it is a pair that many people repulsions. You can choose your favorite type of black tea and use puree instead of fresh peach.

Ingredients		Quantity
물	Water	258g
홍차잎	Black tea leaves	10g
복숭아	Peaches	500g
설탕	Sugar	190g
함수결정포도당	Dextrose	50g
복합안정제	Stabilizer	2g
TOTAL		1,000g

1. 뜨거운 물에 홍차잎을 넣고 우려낸 후 홍차잎을 걸러낸다.
2. 비커에 1과 복숭아를 계량한다.
3. 믹싱볼에 설탕, 함수결정포도당, 복합안정제를 계량한 후 스패출러로 혼합한다.
4. 3을 2에 넣고 핸드블렌더로 믹싱한다.
5. 제조기에 넣고 냉각 교반한다.

● 복숭아 대신 복숭아 퓌레를 사용할 경우 냉장고에서 완전히 해동시킨 후 사용한다.

1. Soak black tea leave in hot water to infuse, then strain the leaves.
2. Weight 1 and peaches in a container.
3. Measure sugar, dextrose, and stabilizer in a mixing bowl and combine with a spatula.
4. Mix 3 with 2 and mix using a hand blender.
5. Pour into the machine and cold-churn the mixture.

● When using peach purée instead of fresh peach, defrost completely in a chiller before use.

당 Sugar	지방 Fat	무지유 고형분 Non-fat milk solids	기타 고형분 Other solids	총 고형분 Total solids	수분 Water
28%	-	-	4.7%	32.7%	67.3%

6 Shine Prosecco Sorbetto

Alcohol

샤인 프로세코 소르베또

산미가 낮고 가벼운 바디감이 특징인 프로세코는 이탈리아 북동부 베네토Veneto 주에서 생산되는 스파클링 와인이다. 달콤한 맛과 과일의 상큼함을 더하기 위해 샤인머스캣을 함께 사용했다. 소르베또 자체로 먹어도 좋고, 소르베또에 프로세코를 곁들여 먹어도 좋다.

Prosecco is a sparkling wine characterized by its low acidity and light body, produced in Veneto, northeastern Italy. I added Shine Muscat to add sweetness and freshness of the fruit. It's good to have sorbetto by itself or with prosecco as well.

Ingredients		Quantity
프로세코	Prosecco	400g
샤인머스캣	Shine Muscat	300g
물	Water	75g
레몬즙	Lemon juice	10g
설탕	Sugar	150g
말토덱스트린	Maltodextrin	40g
이눌린	Inulin	20g
복합안정제	Stabilizer	5g
TOTAL		1,000g

● Contains 4.5% alcohol

1. 비커에 프로세코, 샤인머스켓, 물, 레몬즙을 계량한다.
2. 믹싱볼에 설탕, 말토덱스트린, 이눌린, 복합안정제를 계량한 후 스패출러로 혼합한다.
3. 2를 1에 넣고 핸드블렌더로 믹싱한다.

● 샤인머스켓의 껍질이 씹히는 것이 싫다면 거름망으로 걸러낸다.

4. 제조기에 넣고 냉각 교반한다.

1. Measure prosecco, Shine Muscat, water, and lemon juice in a container.
2. Weight sugar, maltodextrin, inulin, and stabilizer in a mixing bowl and combine with a spatula.
3. Mix 2 with 1 and mix using a hand blender.

● If you don't want the Shine Muscat's skin, stain it through a sieve.

4. Pour into the machine and cold-churn the mixture.

당 Sugar	지방 Fat	무지유 고형분 Non-fat milk solids	기타 고형분 Other solids	총 고형분 Total solids	수분 Water
26%	-	-	2.4%	28.4%	71.6%

7

Strawberry & Mojito Sorbetto

Alcohol

딸기 & 모히또 소르베또

화이트 럼, 라임, 민트가 들어가는 클래식한 모히또에 딸기로 포인트를 주었다. 모히또 특유의 상쾌함과 딸기의 달콤함이 잘 어울리는 메뉴다.

A classic mojito with white rum, lime, and mint is accentuated with strawberries. The freshness of the mojito and the sweetness of the strawberry matches well.

Ingredients		Quantity
화이트 럼	White rum	70g
딸기	Strawberries	400g
물	Water	315g
민트잎	Mint leaves	30g
설탕	Sugar	150g
말토덱스트린	Maltodextrin	40g
이눌린	Inulin	20g
복합안정제	Stabilizer	5g
TOTAL		1,000g

● Contains 2.8% alcohol

1. 비커에 화이트 럼, 딸기, 물, 민트잎을 계량한다.
2. 믹싱볼에 설탕, 말토덱스트린, 이눌린, 복합안정제를 계량한 후 스패출러로 혼합한다.
3. 2를 1에 넣고 핸드블렌더로 믹싱한다.
4. 거름망을 이용해 민트잎을 걸러낸다.
5. 제조기에 넣고 냉각 교반한다.

● 민트잎이 씹히는 것을 선호한다면 걸러내는 작업은 따로 하지 않아도 된다.

1. Measure white rum, strawberries, water, and mint leaves in a container.
2. Weight sugar, maltodextrin, inulin, and stabilizer in a mixing bowl and combine with a spatula.
3. Mix 2 with 1 and mix using a hand blender.
4. Strain the mint leaves through a sieve.
5. Pour into the machine and cold-churn the mixture.

● If you prefer to keep the leaves in the sorbetto, you do not need to strain them.

당 Sugar	지방 Fat	무지유 고형분 Non-fat milk solids	기타 고형분 Other solids	총 고형분 Total solids	수분 Water
23%	-	-	5%	28%	72%

8 Green Tangerine & Makgeolli Sorbetto

Alcohol

청귤 & 막걸리 소르베또

막걸리만을 사용한 소르베또도 좋지만, 여기에서는 청귤을 더해 상큼하면서도 깔끔한 맛으로 마무리되게 만들어보았다. 청량함과 단맛을 표현하기 위해 물 대신 사이다를 사용했다.

Sorbetto made only with makgeolli is good, but I added green tangerine to finish with a tangy and crisp taste. Cider was used instead of water to give refreshing sweetness.

Ingredients		Quantity
막걸리	Makgeolli	420g
청귤즙	Green tangerines juice	200g
사이다	Lemon-lime soda	180g
설탕	Sugar	155g
말토덱스트린	Maltodextrin	40g
복합안정제	Stabilizer	5g
청귤제스트	Zest of 1 or 2 Green tangerines	청귤 1~2개 분량
TOTAL		1,000g

● Contains 2.5% alcohol

1. 비커에 막걸리, 청귤즙, 사이다를 계량한다.
2. 믹싱볼에 설탕, 말토덱스트린, 복합안정제를 계량한 후 스패출러로 혼합한다.
3. 1에 2와 청귤제스트를 넣고 핸드블렌더로 믹싱한다.
4. 제조기에 넣고 냉각 교반한다.

1. Measure makgeolli, green tangerines juice, and Lemon-lime soda in a container.
2. Weight sugar, maltodextrin, and stabilizer in a mixing bowl and combine with a spatula.
3. Mix green tangerine zest, **2** with **1** and mix using a hand blender.
4. Pour into the machine and cold-churn the mixture.

당 Sugar	지방 Fat	무지유 고형분 Non-fat milk solids	기타 고형분 Other solids	총 고형분 Total solids	수분 Water
25%	-	-	2%	27%	73%

Part 8.

Theory for Writing Granita Recipes

그라니따 레시피를 작성하기 위한 이론

Note

- ✓ 레시피를 작성하는 방법은 여러 가지가 있지만, 그 중 가장 쉽게 결과값을 찾을 수 있는 Carpigiani Gelato University의 방법으로 설명하였다.
- ✓ 고형분에서 소수점은 1자리까지 봐주고, 중량에서 소수점은 올리거나 버렸다.

- ✓ There are many ways to write a recipe, but Carpigiani Gelato University's method was explained, which is the easiest to find the result value.
- ✓ The decimal point was counted to one digit, and the decimal point was raised or dropped for the weight.

1 Writing and Understanding Granita Recipes

그라니따 레시피 작성과 이해

그라니따는 이탈리아 남부의 섬, 시칠리아에서 처음 만들어진 콜드 디저트이다. 맛을 내는 주재료와 설탕, 물로만 만들어지므로 젤라또나 소르베또에 비해 얼음 입자가 크고 여름에 더 청량감을 느낄 수 있다. 서빙 및 보관 온도는 -4~-6°C이므로 젤라또와 소르베또가 진열된 쇼케이스에 같이 진열할 수 없다. 이탈리아 젤라떼리아는 여름 시즌에 그라니따만 진열하는 쇼케이스를 사용하거나 그라니따 디스펜서를 사용한다.

Granita is a cold dessert first made in Sicily, an island in southern Italy. It is made only with the main flavoring ingredients, sugar, and water, so the ice particles are larger than gelato or sorbetto, giving a more refreshing taste in summer. Because the serving and storage temperature is -4~-6°C, you cannot display in the same showcase with gelato and sorbetto. Italian gelaterias use showcases that only display granita during the summer season or use granita dispensers.

그라니따 레시피 작성 방법은 젤라또와 소르베또보다 간단하다. 중요하게 결정해야 할 고형분이 '당' 하나이기 때문이다.

Writing a granita recipe is simpler than gelato and sorbetto. It's because the only solids to be determined is 'sugar.'

	당 Sugar
그라니따 Granita	15~22%

과일 그라니따 제조 시 선택한 과일의 당과 몇 %를 넣을 것인지는 소르베또와 동일하게 적용된다.

- 강한 맛 : 15 ~ 30% (예 : 레몬, 라임)
- 중간 맛 : 30 ~ 55% (예 : 딸기, 망고, 바나나 등 80%의 과일이 중간 맛 과일에 속한다.)
- 약한 맛 : 55 ~ 75% (예 : 수박, 오렌지)

과일 향을 한층 더 돋구어주는 레몬즙 역시 소르베또와 동일하게 적용된다.

- 레몬즙 : 0 ~ 2%

When making fruit granita, the sugar and percentage of selected fruit are applied the same as sorbetto.

- Strong taste : 15~30% (e.g. : lemon, lime)
- Medium taste : 30~55% (e.g. : 80% of fruits belong to medium tasting fruits, such as strawberry, mango, and banana.)
- Weak taste : 55~75% (e.g. : watermelon, orange)

Lemon juice, which enhances the fruit flavor, is also used in the same way as sorbetto.

- Lemon juice : 0~2%

2 Strawberry Granita

딸기 그라니따

딸기 그라니따 Strawberry Granita

재료 Ingredients		중량 Weight	당 Sugar
딸기 Strawberry			(8)
레몬즙 Lemon juice			(5)
물 Water			-
설탕 Sugar			(100)
TOTAL	g		
	%		

1. 재료와 재료의 고형분 함량을 적는다.

1. Write down the ingredients and their solids content.

딸기 그라니따 Strawberry Granita

재료 Ingredients		중량 Weight	당 Sugar
딸기 Strawberry			(8)
레몬즙 Lemon juice			(5)
물 Water			-
설탕 Sugar			(100)
TOTAL	g	1,000 **(1)**	200 **(3)**
	%		20% **(2)**

2. 만들고자 하는 총 합계 중량 **(1)**과 최종 당 비율 **(2)** 그에 따른 당값 **(3)**을 설정한다.

2. Set the total weight to make **(1)**, the final sugar ratio **(2)**, and the sugar value **(3)** accordingly.

딸기 그라니따 Strawberry Granita

재료 Ingredients		중량 Weight	당 Sugar
딸기 Strawberry		500 **(1)**	(8) 40 **(2)**
레몬즙 Lemon juice			(5)
물 Water			-
설탕 Sugar			(100)
TOTAL	g	1,000	200
	%		20%

3. 과일을 몇 %로 넣을지 정하고 당을 계산한다.

딸기는 중간 맛의 강도를 가진 과일이니 35~55% 가이드를 가진다.

이 레시피는 50%의 비율로 계산하였다.

딸기 (당 8%)

중량 : 총 중량 1,000 × 딸기 50% = 500 **(1)**

당 : 중량 200 × 당 8% = 40 **(2)**

3. Decide what percentage of fruit to add and calculate the sugar.

Strawberries are fruits with medium taste intensity, and belong to the 35~55% guide.

The recipe was calculated at a ratio of 50%.

Strawberry (8% sugar)

Weight : Total weight 1,000 × 50% strawberries = 500 **(1)**

Sugar : Weight 200 × 8% sugar = 40 **(2)**

딸기 그라니따 Strawberry Granita

재료 Ingredients		중량 Weight	당 Sugar
딸기 Strawberry		500	(8) 40
레몬즙 Lemon juice		10 **(1)**	(5) 0.5 **(2)**
물 Water			-
설탕 Sugar			(100)
TOTAL	g	1,000	200
	%		20%

4. 레몬즙을 몇 %로 넣을지 정하고 당을 계산한다.

그라니따에 레몬즙은 0~2% 들어간다.

이 레시피는 1%의 비율로 계산하였다.

레몬 (당 5%)

중량 : 총 중량 1,000 × 레몬 1% = 10 **(1)**

당 : 중량 10 × 당 5% = 0.5 **(2)**

4. Decide what percentage of lemon juice to add and calculate the sugar.

Granita contains 0~2% lemon juice.

The recipe was calculated at a ratio of 1%.

Lemon (5% sugar)

Weight : Total weight 1,000 × 1% lemon = 10 **(1)**

Sugar : Weight 10 × 5% sugar = 0.5 **(2)**

딸기 그라니따 Strawberry Granita

재료 Ingredients		중량 Weight	당 Sugar
딸기 Strawberry		500	(8) 40
레몬즙 Lemon juice		10	(5) 0.5
물 Water			-
설탕 Sugar		160 **(2)**	(100) 159.5 **(1)**
TOTAL	g	1,000	200
	%		20%

5. 당의 총 무게에서 딸기와 레몬즙의 당을 뺀 값이 설탕의 당량이 된다.

설탕 (당 100%)

당 : 최종 당 200 – 딸기 당 40 – 레몬즙 당 0.5 = 159.5

설탕은 100% 고형분이기에 당량 **(1)**, 중량 **(2)**이 동일하다.

5. The value obtained by subtracting the sugar in strawberry and lemon juice from the total weight of sugar is sugar weight.

Sugar (100% sugar)

Sugar : Final sugar 200 - Sugar in strawberry 40 - Sugar in lemon juice 0.5 = 159.5

Since sugar is 100% solids, the sugar weight **(1)** and the weight of ingredient **(2)** are the same.

딸기 그라니따 Strawberry Granita

재료 Ingredients		중량 Weight	당 Sugar
딸기 Strawberry		500	(8) 40
레몬즙 Lemon juice		10	(5) 0.5
물 Water		330 **(1)**	-
설탕 Sugar		160	(100) 159.5
TOTAL	g	1,000	200
	%		20%

6. 마지막 물은 총 중량에서 물을 제외한 모든 재료의 중량을 뺀 값이다.

물 (수분 100%)

중량 : 총 중량 1,000 – 딸기 500 – 레몬즙 10 – 설탕 160 = 330 **(1)**

6. As for the water, it is the total weight minus the weight of all ingredients except water.

Water (100% water)

Weight : Total weight 1,000 - Strawberry 500 - Lemon juice 10 - Sugar 160 = 330 **(1)**

Part 9.

Granita Recipes

그라니따 레시피

1 Lemon Granita

레몬 그라니따

레몬의 상큼함 그 자체를 즐길 수 있는 레몬 그라니따. 기호에 따라 좋아하는 허브를 더하거나 장식해 향을 더해보는 것도 좋다.

Lemon granita lets you taste refreshing lemon itself. You can add or decorate with your favorite herbs to your taste to add aroma.

Ingredients		Quantity
레몬즙	Lemon juice	200g
물	Water	610g
설탕	Sugar	190g
TOTAL		1,000g

1. 비커에 모든 재료를 계량한다.
2. 핸드블렌더로 설탕 입자가 느껴지지 않을때까지 혼합한다.
3. 레몬 제스트를 넣는다.
4. 제조기에 넣고 냉각 교반한다.

1. Measure all the ingredients in a container.
2. Mix with a hand blender until all the sugar is dissolved.
3. Add lemon zest.
4. Pour into the machine and cold-churn the mixture.

2

Caffé Granita

커피 그라니따

진한 커피맛이 느껴지는 그라니따에 달콤한 생크림을 더한 메뉴. 생크림을 그라니따 아래에 담거나 위에 담을 수 있는데, 따로따로 먹다가 함께 비벼 먹어도 좋다.

It is a menu with sweetened whipped cream added to a robust coffee-flavored granita. You can put the cream under or over the granite to eat separately or mixed together.

커피 그라니따
Caffé granita

Ingredients		Quantity
모카포트로 내린 커피	Coffee brewed with the moka pot	380g
인스턴트 커피가루	Instant coffee powder	5g
물	Water	415g
설탕	Sugar	200g
TOTAL		1,000g

1. 모카포트로 커피를 내린다.
● 모카포트로 내린 커피 대신 아메리카노나 더치커피로 대체할 수 있다.
2. 비커에 모든 재료를 넣는다.
3. 핸드블렌더로 설탕 입자가 느껴지지 않을 때까지 혼합한다.
4. 제조기에 넣고 냉각 교반한다.

1. Brew coffee with the moka pot.
● The brewed coffee can be replaced with Americano or Dutch coffee.
2. Put all the ingredients in a container.
3. Whip until all the sugar is dissolved.
4. Pour into the machine and cold-churn the mixture.

토핑 생크림
Whipped cream topping

Ingredients		Quantity
생크림	Cream (38% fat)	435g
설탕	Sugar	65g
TOTAL		500g

1. 믹싱볼에 생크림을 넣고 설탕을 3번 정도 나눠 넣어가며 휘핑한다.
2. 완성된 토핑 생크림은 커피 그라니따 위에 올린다.

1. Put whipping cream in a mixing bowl and whip while adding sugar in three parts.
2. Top the whipped cream on the coffee granita.

3 Almond Granita

아몬드 그라니따

레몬 그라니따와 커피 그라니따와 함께 시칠리아에서 가장 유명한 아몬드 그라니따. 시칠리아에서는 아침이나 점심에 그라니따를 브리오슈와 함께 곁들여 먹기도 한다. 여기에서는 직접 만든 아몬드 반죽으로 아몬드의 풍미와 밀도감을 더했다.

Almond granite is Sicily's most famous, along with lemon granita and coffee granite. In Sicily, granite is often served with brioche for breakfast or lunch. I added homemade almond paste to add the flavor and density of almonds.

아몬드 반죽*
Almond dough

Ingredients		Quantity
설탕A	Sugar A	80g
물	Water	80g
껍질 벗긴 아몬드	Blanched almonds	460g
설탕B	Sugar B	380g
TOTAL		1,000g

1. 냄비에 설탕A와 물을 넣고 105°C까지 끓여 설탕 시럽을 만든다.
2. 푸드프로세서에 껍질 벗긴 아몬드, 설탕B를 넣고 갈아준다.
3. 2를 믹싱볼에 옮긴다.
4. 3에 1을 부어가면서 믹싱해 반죽으로 만든다.

● 아몬드 반죽은 냉장고에서 2주간 보관하며 사용할 수 있다.

1. Boil sugar A and water in a pot to 105 to make a sugar syrup.
2. Grind blanched almonds and sugar B in a food processor.
3. Transfer 2 into a mixing bowl.
4. Gradually pour 1 into 3 while mixing to make a dough.

● Almond dough can be stored in a refrigerator for two weeks.

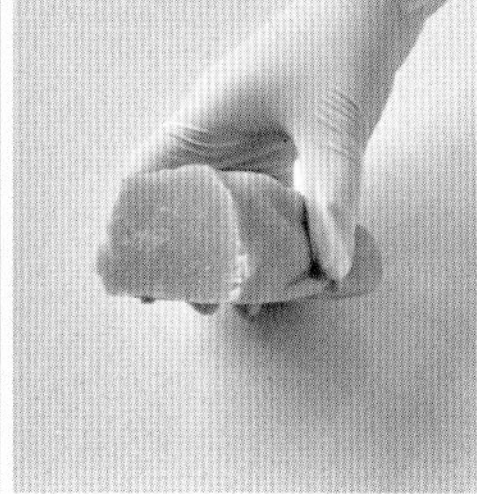

아몬드 그라니따
Almond granita

Ingredients		Quantity
아몬드 반죽*	Almond dough*	330g
물	Water	670g
TOTAL		1,000g

1. 비커에 모든 재료를 계량한다.
2. 휘퍼로 완전히 혼합한다.
3. 제조기에 넣고 냉각 교반한다.

1. Measure all the ingredients in a container.
2. Whip to combine completely.
3. Pour into the machine and cold-churn the mixture.

Part 10.

Cold Dessert Recipes

콜드 디저트 레시피

1 Matcha Gelato & Sweet Red Bean Monaka

말차 젤라또 & 팥 모나카

바삭한 과자 사이에 충전물을 샌딩해 만드는 모나카는 젤라또와 쉽게 조합할 수 있는 디저트 중 하나다. 모나카의 바삭함을 더 오래 유지하기 위해 피 안쪽에 화이트초콜릿을 얇게 발라 쉽게 눅눅해지지 않도록 했다. 여기에서는 녹차 젤라또와 팥앙금을 샌딩했지만 다양한 맛의 젤라또로 응용할 수 있다.

Monaka is made by sandwiching the filling between crispy cookie shells, and it is one of the desserts you can easily combine with gelato. In order to keep the crispness of monaka longer, I applied a thin layer of white chocolate to the inside of the shell to prevent it from turning soggy quickly. I used Matcha gelato and sweet red beans for filling, but any flavor of gelato can be used.

말차 젤라또
Matcha gelato

Ingredients		Quantity
우유	Milk	630g
생크림	Cream (38% fat)	120g
물엿	Glucose syrup	30g
탈지분유	Skim milk powder	30g
설탕	Sugar	125g
함수결정포도당	Dextrose	40g
복합안정제	Stabilizer	5g
말차가루	Matcha powder	20g
TOTAL		1,000g

1. 냄비에 우유, 생크림, 물엿을 넣고 40°C로 가열한다.
2. 믹싱볼에 탈지분유, 설탕, 함수결정포도당, 복합안정제, 말차가루를 계량한 후 스패출러로 혼합한다.
3. 2를 1에 넣고 65~85°C로 가열한다.
4. 4°C로 냉각 후 0-12시간 숙성한다.
5. 제조기에 넣고 냉각 교반한다.

1. Heat milk, cream, and glucose syrup in a pot to 40°C.
2. Measure skim milk powder, sugar, dextrose, stabilizer, and matcha powder in a mixing bowl and combine with a spatula.
3. Add 2 into 1 and heat to 65~85°C.
4. Cool to 4°C and let rest for 0~12 hours.
5. Pour into the machine and cold-churn the mixture.

기타
Other

Ingredients	
모나카(시판)	Monaka (store-bought)
팥앙금(시판)	Sweet red bean paste (store-bought)
화이트초콜릿 적당량	White chocolate QS

몽타주
Montage

1. 모나가 안쪽에 녹인 화이드초콜릿을 한 겹 바른다.
2. 1에 말차 젤라또를 한 스쿱 올린다.
3. 팥앙금을 올린 후 모나카를 덮는다.

1. Brush a layer of melted white chocolate on the inside of the monaka.
2. Put a scoop of matcha gelato in the shell.
3. Top with sweet red bean paste and cover with the remaining shell.

2 Bottle Gelato

보틀 젤라또

이탈리아 젤라떼리아에서는 젤라또를 콘이나 컵으로만 판매하지 않고 여러 가지 재료를 보틀에 층층이 담아 판매하기도 한다. 맛도 맛이지만 켜켜이 쌓인 다양한 재료들이 시각적인 매력을 주어 선물용 패키지로 만들어 판매하기에도 좋다. -18°C 이하의 스탠드형 냉동고에 진열해 판매한다.

Italian Gelateria sells gelato not only in cones or cups but also in layers with different ingredients in bottles. Of course, it's delicious, but various ingredients stacked in layers are visually appealing, making it suitable to sell as a gift package. Showcased and sold in upright freezers below -18°C.

비스퀴 조콩드
Biscuit joconde

Ingredients		Quantity
슈거파우더	Powdered sugar	35g
박력분	Cake flour	30g
아몬드가루	Almond powder	100g
우유	Milk	20g
노른자	Egg yolks	75g
버터	Butter	75g
흰자	Egg whites	120g
설탕	Sugar	45g
TOTAL		500g

1. 슈거파우더, 박력분, 아몬드가루를 체 친다.
2. 푸드프로세서에 1과 우유, 노른자를 넣고 믹싱한다.
3. 2에 녹인 버터를 넣고 섞는다.
4. 볼에 흰자와 설탕을 넣고 휘핑해 머랭을 만든다.
5. 3에 4를 넣고 섞는다.
6. 테프론시트를 깐 철판에 평평하게 펼친다.
7. 160°C로 예열된 오븐에서 15분간 굽는다.

1. Sift powdered sugar, cake flour, and almond powder.
2. Mix 1 with milk and egg yolks in a food processor.
3. Add melted butter to 2 and mix.
4. Whip egg whites and sugar in a mixing bowl to make meringue.
5. Add 4 with 3 to combine.
6. Spread evenly on a baking tray lined with a Teflon sheet.
7. Bake for 15 minutes in an oven preheated to 160°C.

캐러멜 소스*
Caramel sauce

Ingredients		Quantity
설탕	Sugar	450g
함수결정포도당	Dextrose	50g
생크림	Cream (38% fat)	400g
버터	Butter	100g
TOTAL		1,000g

1. 물기 없는 냄비에 설탕과 함수결정포도당을 계량한 후 170°C까지 가열하며 설탕을 녹인다.
2. 설탕이 완전히 녹으면 불을 끈 후 생크림을 천천히 부으며 섞어준다.
3. 생크림이 혼합되면 버터를 넣고 녹여준다.

● 냉장고에서 15일간 보관하며 사용 가능하다.

1. Measure sugar and dextrose in a dry pot. Heat to 170°C and dissolve the sugar.
2. When the sugar is completely dissolved, remove from heat and slowly add cream while stirring.
3. When the cream is combined, add butter.

● Can be stored in a refrigerator for 15 days.

솔티드 캐러멜 젤라또
Salted caramel gelato

Ingredients		Quantity
우유	Milk	555g
생크림	Cream (38% fat)	50g
탈지분유	Skim milk powder	45g
설탕	Sugar	30g
함수결정포도당	Dextrose	10g
복합안정제	Stabilizer	5g
소금	Salt	5g
캐러멜 소스*	Caramel sauce*	300g
TOTAL		1,000g

1. 냄비에 우유, 생크림을 넣고 40°C로 가열한다.
2. 믹싱볼에 탈지분유, 설탕, 함수결정포도당, 복합안정제, 소금을 계량한 후 스패출러로 혼합한다.
3. 2를 1에 넣고 65~85°C로 가열한다.
4. 4°C로 냉각한다.
5. 캐러멜 소스를 넣고 핸드블렌더로 혼합한다.
6. 0~12시간 숙성한다.
7. 제조기에 넣고 냉각 교반한다.

1. Heat milk and cream in a pot to 40°C.
2. Measure skim milk powder, sugar, dextrose, stabilizer, and salt in a mixing bowl and combine with a spatula.
3. Add 2 into 1 and heat to 65~85°C.
4. Cool to 4°C.
5. Add caramel sauce and mix with a hand blender.
6. Let rest for 0~12 hours.
7. Pour into the machine and cold-churn the mixture.

아몬드 소르베또
Almond sorbetto

Ingredients		Quantity
물	Water	580g
아몬드 반죽	Almond dough	300g
설탕	Sugar	90g
함수결정포도당	Dextrose	25g
복합안정제	Stabilizer	5g
TOTAL		1,000g

1. 비커에 모든 재료를 계량한다.

● 아몬드 반죽은 아몬드 그라니따(245p)의 아몬드 반죽 공정과 동일하다.

2. 핸드 블렌더로 믹싱한다.
3. 제조기에 넣고 냉각 교반한다.

1. Measure all the ingredients in a container.

● Refer to "Almond Granita (p.245)" for the almond dough.

2. Mix using a hand blender.
3. Pour into the machine and cold-churn the mixture.

허니콤
Honeycomb

Ingredients		Quantity
설탕	Sugar	168g
꿀	Honey	75g
물	Water	45g
베이킹소다	Baking soda	12g
TOTAL		1,000g

1. 냄비에 설탕, 꿀, 물을 계량한 후 160°C까지 가열한다.
2. 설탕이 완전히 녹으면 불에서 내린 후 베이킹소다를 넣으며 섞어준다.

● 내용물이 부풀어오르므로 깊은 냄비를 사용하는 것이 좋다.

3. 테프론시트에 부어 식힌 후 조각 내 사용한다.

1. Measure sugar, honey, and water in a pot and heat to 160°C.
2. When the sugar is completely dissolved, remove from the heat and mix with baking soda.

● It's best to use a deep pot as the contents will rise.

3. Pour onto a Teflon sheet, let cool, and break into pieces to use.

바닐라 젤라또
Vanilla gelato

Ingredients		Quantity
우유	Milk	640g
생크림	Cream (38% fat)	140g
물엿	Glucose syrup	30g
바닐라빈	Vanilla beans	2개
탈지분유	Skim milk powder	30g
설탕	Sugar	130g
함수결정포도당	Dextrose	25g
복합안정제	Stabilizer	5g
TOTAL		1,000g

1. 냄비에 우유, 생크림, 물엿, 바닐라빈을 넣고 40°C로 가열한다.
2. 믹싱볼에 탈지분유, 설탕, 함수결정포도당, 복합안정제를 계량한 후 스패출러로 혼합한다.
3. 2를 1에 넣고 65~85°C로 가열한다.
4. 4°C로 냉각한다.
5. 0~12시간 숙성한다.
6. 제조기에 넣고 냉각 교반한다.

1. Heat milk, cream, glucose syrup, and vanilla beans in a pot to 40°C.
2. Measure skim milk powder, sugar, dextrose, and stabilizer in a mixing bowl and combine with a spatula.
3. Add 2 into 1 and heat to 65~85°C.
4. Cool to 4°C.
5. Let rest for 0~12 hours.
6. Pour into the machine and cold-churn the mixture.

기타
Other

Ingredients	
아몬드 분태	Chopped almonds

몽타주
Montage

1. 보틀에 비스퀴 조콩드를 넣는다.

● 비스퀴 조콩드는 보틀의 크기에 맞춰 잘라 사용한다.

2. 1 위에 솔티드 캐러멜 젤라또 - 아몬드 소르베또 - 허니콤 - 바닐라 젤라또 순서로 넣는다.
3. 기호에 따라 캐러멜 소스와 아몬드 분태를 뿌린다.

1. Place a joconde biscuit in the bottle.

● Cut the biscuit according to the size of the bottle.

2. Put in order on top of the biscuit; Salted caramel gelato - Almond sorbetto - Honeycomb - Vanilla gelato.
3. Sprinkle with caramel sauce and chopped almond according to taste.

3 Gelatini

젤라띠니

젤라띠니는 쉽고 간단하게 만들 수 있어 이탈리아 젤라떼리아에서 많이 볼 수 있는 디저트 중 하나다. 미니 콘에 원하는 맛의 젤라또를 한 스쿱 올린 후 초콜릿으로 디핑하면 완성이다. 기호에 따라 견과류 등의 부재료를 토핑해도 좋다. 이탈리아에서는 낱개로도, 무게를 재어 판매하기도 한다. -18°C 이하의 스탠드형 냉동고에 진열해 판매한다.

Gelatini is one of the most popular desserts in Italian Gelateria because it's easy and straightforward. Add a scoop of gelato of your choice to a mini cone, then dip it in chocolate to finish. You may top it with additional ingredients, such as nuts, to your taste. In Italy, it's sold individually or by weight. Display and sell in upright freezers below -18°C.

미니콘에 젤라또를 담아 냉각시킨 후 템퍼링한 커버추어 초콜릿 또는 녹인 코팅 초콜릿에 담갔다 빼 굳힌다.

After freezing the gelato in mini cones, dip them in tempered couverture chocolate or melted coating chocolate to set.

4

Gelato Shake & Sorbetto Ade

젤라또 셰이크 & 소르베또 에이드

젤라또에 우유를 더하고 믹싱해 젤라또 셰이크를 만들거나, 소르베또에 사이다를 더해 소르베또 에이드로 만들 수 있다. 이미 만들어져 있는 젤라또나 소르베또를 이용할 수 있으므로 젤라떼리아에서 간편하게 만들어 여름 시즌 메뉴로 판매하기 좋다.

Add milk to gelato and mix to make a gelato shake or mix with lemon-lime soda to sorbetto to make sorbetto ade. Since you can use ready-made gelato or sorbetto, it's easy to make in the Gelateria and sell them as a summer season menu.

INDEX 젤라또를 만드는 데 사용되는 시판 재료들

베이스50 Base50

FABBRI
NEVEPANN 50 C/F

MEC3
50 BASE NAPURE

MEC3
PANNA BASE

PREGEL
BASE PANNA MILK 50

- FABBRI 수입사: (주)지이피트레이딩
- MEC3, GIUSO 수입사: (주)에프앤지코리아
- PreGel 수입사: (주)한아통상
- boiron 수입사: (주)제원인터내쇼날
- adamance 수입사: (주)베이크플러스

복합안정제 Stabilizer

FABBRI
Neutro SP

MEC3
NEUTRANIN

과일 퓌레 Fruits purée

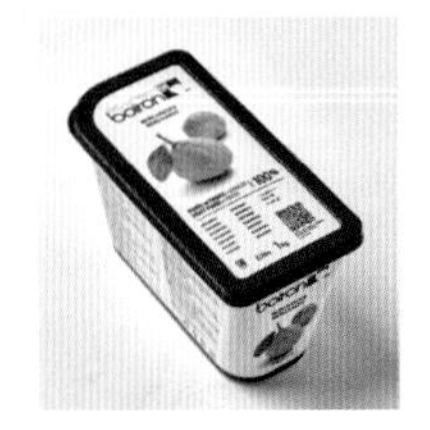

boiron
BERGAMOT

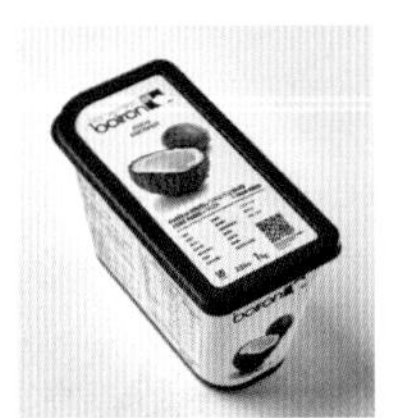

boiron
COCONUT

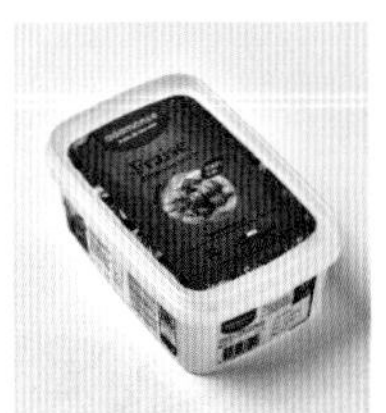

adamance
Fraise

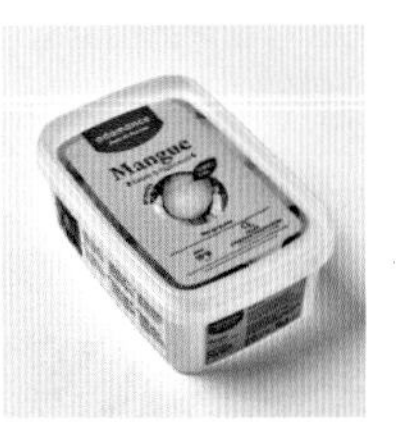

adamance
Mangue

견과류 페이스트 Nut paste

FABBRI
MANDORLA • ALMOND

FABBRI
NOCCIOLA PIEMONTE IGP

GIUSO
PISTACCHIO SICILIA INTEGRALE GOLD

MEC3
NOCCIOLA-LA-LA SCURA

MEC3
PASTA PISTACCHIO 100% SICILIA

PreGel
PISTACHIO PURO N

Commercially Available Ingredients used to Make Gelato

과일 페이스 Fruits paste

FABBRI
SIMPLE FRAGOLA

PreGel
BILBERRY N

당 페이스 Sugar paste

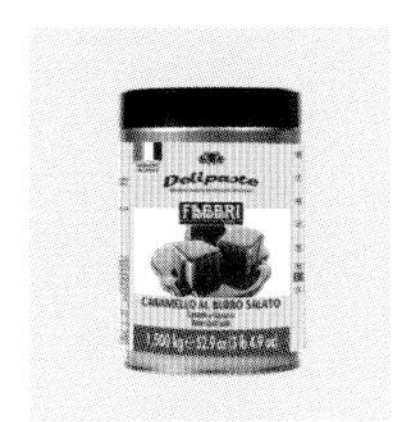

FABBRI
CARAMELLO AL BURRO SALAT

FABBRI
MARRONS GLACES

FABBRI
MENTA BIANCA • WHITE MINT

MEC3
Pasta Menta Napure

MEC3
Pasta Menta

MEC3
Pasta French vanilla

소스 Sauce

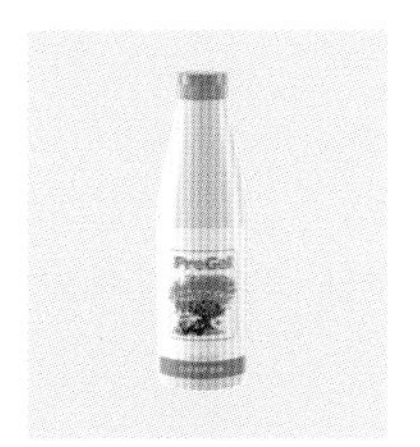

PreGel
CHERRY BON

FABBRI
TOPPING AMARENA PEZZI

FABBRI
TOPPING NOCCIOLA & CIOCCOLATO

MEC3
TOPPING LAMPONE

MEC3
STRACCIATELLA DARK

MEC3
STRACCIATELLA PINK

토핑 Topping

FABBRI
VARIEGATO HONEY & QUINOA

FABBRI
VARIEGATO AMARENA N secchiello

FABBRI
VARIEGATO AMARENA Dall'alto

이 페이지는 젤라또, 소르베또, 그라니따 레시피를 작성을 위한 연습 페이지입니다. 필요한 만큼 복사해서 사용하세요.

This page is to practice writing gelato, sorbetto, and granita recipes. Copy and use as many as you need.

재료 Ingredients		중량 (g) Weight	당 (%) Sugar	지방 (%) Fat	무지유 고형분 (%) Non-fat milk solids	기타 고형분 (%) Other solids	총 고형분 (%) Total solids
TOTAL	g						
	%						

재료 Ingredients		중량 (g) Weight	당 (%) Sugar	지방 (%) Fat	무지유 고형분 (%) Non-fat milk solids	기타 고형분 (%) Other solids	총 고형분 (%) Total solids
TOTAL	g						
	%						

73~74p와 동일한 '젤라또에 사용되는 재료들의 평균 고형분 함량' 표입니다. 책과 함께 나란히 두고 보면 더 편리하게 계산할 수 있습니다.

This is the same 'Average solids content of ingredients used in gelato' table as on pages 73~74. You can calculate more conveniently if you put them side by side.

		당 Sugar	지방 Fat	무지유 고형분 Non-fat milk solids	기타 고형분 Other solids	총 고형분 Total solids	수분 Water
유제품 Dairy product	일반 우유 Whole milk	-	3.5%	9%	-	12.5%	87.5%
	저지방 우유 Low-fat milk	-	1.8%	9%	-	10.8%	89.2%
	무지방 우유 Non-fat milk	-	0.2%	9%	-	9.2%	90.8%
	생크림 (35% 지방) Cream (35% fat)	-	35%	5.8%	-	40.8%	59.2%
	생크림 (38% 지방) Cream (38% fat)	-	38%	5.6%	-	43.6%	56.4%
	전지분유 Whole milk powder	-	26%	71%	-	97%	3%
	탈지분유 Skim milk powder	-	1%	95%	-	96%	4%
	버터 Butter	-	84%	-	-	84%	16%
	가당 연유 Sweetened condensed milk	43%	9%	24%	-	76%	24%
	무가당 연유 Unsweetened condensed milk	-	8~9%	18~24%	-	26~33%	74~67%
	요거트 Yogurt	-	4%	9.5%	-	13.5%	86.5%
	리코타 Ricotta	-	13%	12%	-	25%	75%
	마스카르포네 Mascarpone	-	47%	8.5%	-	55.5%	44.5%
	크림치즈 Cream cheese	-	31%	10%	-	41%	59%
당류 Sugars	설탕 Sugar	100%	-	-	-	100%	-
	함수결정포도당 Dextrose	92%	-	-	-	92%	8%
	물엿 Glucose syrup	80%	-	-	-	80%	20%
	글루코스 시럽 파우더 Glucose syrup powder	96%	-	-	-	96%	4%
	전화당 Inverted sugar	75%	-	-	-	75%	25%
	꿀 Honey	80%	-	-	-	80%	20%
	트레할로스 Trehalose	100%	-	-	-	100%	-
	말토덱스트린 Maltodextrin	96%	-	-	-	96%	4%
	과당 Fructose	100%	-	-	-	100%	-
	메이플 시럽 Maple syrup	67%	-	-	-	67%	33%
	이눌린 Inulin	-	-	-	-	94%	4%
원재료 및 가공 제품 Raw material & processed products	달걀 Eggs	-	14%	-	11%	25%	75%
	노른자 Egg yolk	-	30%	-	18%	48%	52%
	흰자 Egg white	-	-	-	15%	15%	85%
	카카오 파우더 (10~12%) Cocoa powder (10~12% fat)	-	11%	-	84%	95%	5%
	카카오 파우더 (22~24%) Cocoa powder (22~24% fat)	-	23%	-	72%	95%	5%
	카카오매스 Cocoa mass	-	55%	-	44%	99%	1%
	화이트초콜릿 White chocolate	55%	20%	15%	10%	100%	-
	밀크초콜릿 Milk chocolate	45%	36%	12.5%	6.5%	100%	-
	다크초콜릿 (70%) Dark chocolate (70%)	30%	40%	-	30%	100%	-
	땅콩 Peanut	3%	50%	-	44%	97%	3%
	아몬드 Almond	4.5%	55%	-	40.5%	100%	-
	잣 Pine nut	4%	50%	-	42%	96%	4%
	호두 Walnut	4%	68%	-	28%	100%	-
	헤이즐넛 페이스트 Hazelnut paste	-	65%	-	35%	100%	-
	피스타치오 페이스트 Pistachio paste	-	55%	-	45%	100%	-
	베이스50 크림 (젤라또용) Base50 Cream (for gelato) 베이스50 cc Base50 cc	40%	-	40%	14%	94%	6%
	베이스50 과일 (소르베또용) Base50 Fruit (for sorbetto) 베이스50 ff Base50 ff	80%	-	-	14%	94%	6%
	복합안정제 Stabilizer	-	-	-	100%	100%	-

Gelato

‖ SORBETTO, GRANITA, COLD DESSERT

First edition published	September 1, 2023
Second edition published	July 5, 2024
Author	Yoo Siyeon
Translated by	Kim Eunice
Publisher	Bak Yunseon
Published by	THETABLE Inc.
Plan & Edit	Bak Yunseon
Design	Kim Bora
Photograph	Jo Wonseok
Stylist	Lee Hwayoung
Sales/Marketing	Kim Namkwon, Cho Yonghoon, Moon Seongbin
Management support	Kim Hyoseon, Lee Jungmin
Address	122, Jomaru-ro 385beon-gil, Bucheon-si, Gyeonggi-do, Republic of Korea
Website	www.icoxpublish.com
Instagram	@thetable_book
E-mail	thetable_book@naver.com
Phone	82-32-674-5685
Registration date	August 4, 2022
Registration number	386-2022-000050
ISBN	979-11-92855-01-1 (13590)

—

The foreign words in this book follow the Regulation of Loan orthography set by the National Institute of the Korean Language, but some words are written close to Italian pronunciation.